ROSEVILLE PUBLIC LIBRARY
5115
P9-DGB-371

WHY GOD WON'T GO AWAY

Also by Eugene d'Aquili and Andrew Newberg

The Mystical Mind: Probing the Biology of Religious Experience

WHY GOD WON'T GO AWAY

Brain Science
and the Biology of Belief

ANDREW NEWBERG, M.D.

EUGENE D'AQUILI, M.D., Ph.D.

and VINCE RAUSE

BALLANTINE BOOKS • NEW YORK

A Ballantine Book
Published by The Ballantine Publishing Group

Grateful acknowledgment is made to the following for permission to reprint
previously published material:

Harcourt, Inc.: "Footnote to All Prayers" from *Poems* by C. S. Lewis.
Copyright © 1964 by the Executors of the Estate of C. S. Lewis and renewed
1992 by C. S. Lewis Pte Ltd. Reprinted by permission of Harcourt, Inc.

The McGraw-Hill Companies: three figures from *Principles of Neural Science*,
3rd edition, by Kandel and Schwartz, 1992. Reprinted by permission of
The McGraw-Hill Companies.

New World Library: excerpts from *The Mystic Heart* by Wayne Teasdale.
Copyright © 1999 by Wayne Teasdale. Reprinted by permission of New World
Library, Novato, CA, www. newworldlibrary.com.

Random House, Inc.: excerpts from *A History of God* by Karen Armstrong.
Copyright © 1993 by Karen Armstrong. Reprinted by permission of
Alfred A. Knopf, a division of Random House, Inc.

www.randomhouse.com/BB/

Illustrations by Judith Cummins

Cataloging-in-Publication Data can be obtained
from the publisher upon request.

ISBN 0-345-44033-1

Text design by Holly Johnson

Manufactured in the United States of America

First Edition: April 2001

10 9 8 7 6 5 4 3 2 1

To our families

CONTENTS

1. A Photograph of God?: An Introduction to the Biology of Belief 1
2. Brain Machinery: The Science of Perception 11
3. Brain Architecture: How the Brain Makes the Mind 35
4. Myth-making: The Compulsion to Create Stories and Beliefs 54
5. Ritual: The Physical Manifestation of Meaning 77
6. Mysticism: The Biology of Transcendence 98
7. The Origins of Religion: The Persistence of a Good Idea 128
8. Realer Than Real: The Mind in Search of Absolutes 142
9. Why God Won't Go Away: The Metaphor of God and the Mythology of Science 157

Notes 173
References 201
Index 217

1

A PHOTOGRAPH OF GOD?

An Introduction to the Biology of Belief

In a small, dark room at the lab of a large university hospital, a young man named Robert lights candles and a stick of jasmine incense; he then settles to the floor and folds his legs easily into the lotus position. A devout Buddhist and accomplished practitioner of Tibetan meditation, Robert is about to begin another meditative voyage inward. As always, his goal is to quiet the constant chatter of the conscious mind and lose himself in the deeper, simpler reality within. It's a journey he's made a thousand times before, but this time, as he drifts off into that inner spiritual reality—as the material world around him recedes like a fading dream—he remains tethered to the physical here and now by a length of common cotton twine.

One end of that twine lies in a loose coil at Robert's side. The other end runs beneath a closed laboratory door and into an adjoining room, where I sit, beside my friend and longtime research partner Dr. Eugene d'Aquili, with the twine wrapped around my finger. Gene and I are waiting for Robert to tug on the twine, which will be our signal that his meditative state is approaching its transcendent peak. It is this peak moment of spiritual intensity that interests us.[1]

For years, Gene and I have been studying the relationship between religious experience and brain function, and we hope that by monitoring Robert's brain activity at the most intense and mystical moments of his meditation, we might shed some light on the mysterious connection between human consciousness and the persistent and peculiarly human longing to connect with something larger than ourselves.

In earlier conversations, Robert has struggled to describe for us how he feels as his meditation progresses toward this spiritual peak. First, he says, his conscious mind quiets, allowing a deeper, simpler part of himself to emerge. Robert believes that this inner self is the truest part of who he is, the part that never changes. For Robert, this inner self is not a metaphor or an attitude; it is literal, constant, and real. It is what remains when worries, fears, desires, and all other preoccupations of the conscious mind are stripped away. He considers this inner self the very essence of his being. If pressed, he might even call it his soul.[2]

Whatever Robert calls this deeper consciousness, he claims that when it emerges during those moments of meditation when he is most completely absorbed in looking inward, he suddenly understands that his inner self is not an isolated entity, but that he is inextricably connected to all of creation. Yet when he tries to put this intensely personal insight into words, he finds himself falling back on familiar clichés that have been employed for centuries to express the elusive nature of spiritual experience. "There's a sense of timelessness and infinity," he might say. "It feels like I am part of everyone and everything in existence."[3]

To the traditional scientific mind, of course, these terms are useless. Science concerns itself with that which can be weighed, counted, calculated, and measured—anything that can't be verified by objective observation simply can't be called scientific. Although individual scientists might be personally intrigued by Robert's experience, as professionals they'd likely dismiss his comments as too personal and speculative to signify anything concrete in the physical world.[4]

Years of research, however, have led Gene and me to believe that experiences like Robert's are real, and can be measured and verified by solid science.[5] That's exactly why I'm huddling, beside Gene, in this cramped examination room, holding kite string between my fingers: I'm waiting for Robert's moment of mystical transcendence to arrive, because I intend to take its picture.[6]

We wait one hour, while Robert meditates. Then I feel a gentle jerk on the twine. This is my cue to inject a radioactive material into a long intravenous line that also runs into Robert's room, and into a vein in his left arm. We wait a few moments more for Robert to end his meditation, then we whisk him off to a room in the hospital's Nuclear Medicine Department, where a massive, state-of-the-art SPECT camera awaits. In moments, Robert is reclining on a metal table, the camera's three large crystal heads orbiting his skull with a precise, robotic whir.

The SPECT camera (the acronym stands for single photon emission computed tomography) is a high-tech imaging tool that detects radioactive emissions.[7] The SPECT camera scans inside Robert's head by detecting the location of the radioactive tracer we injected when Robert tugged on the string. Because the tracer is carried by blood flow, and because this particular tracer locks almost immediately into brain cells and remains there for hours, the SPECT scans of Robert's head will give us an accurate freeze-frame of blood flow patterns in Robert's brain just moments after injection—at the high point of his meditative climax.

Increased blood flow to a given part of the brain correlates with heightened activity in that particular area, and vice versa.[8] Since we have a good idea of the specific functions that are performed by various brain regions, we expect the SPECT images to tell us a lot

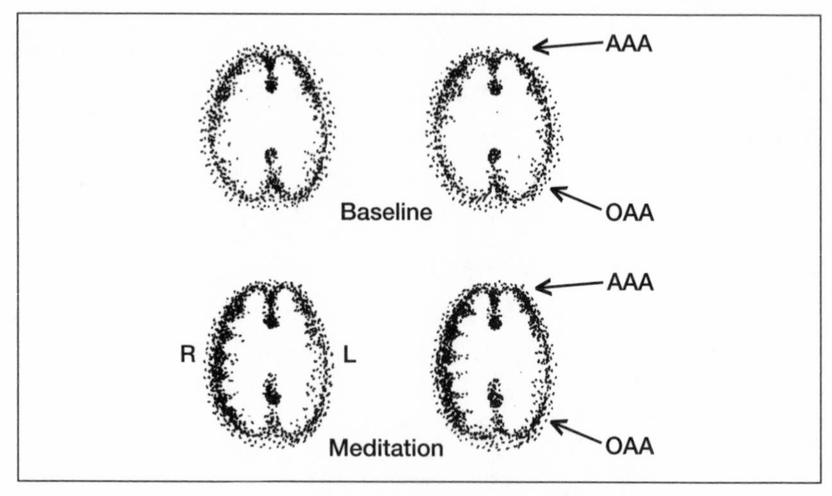

Figure 1.1: The top row of images shows the meditator's brain at rest and indicates an even distribution of activity throughout the brain. (The top of the image is the front of the brain and part of the attention association area, or AAA, while the bottom of the image is part of the orientation association area, or OAA.) The bottom row of images shows the brain during meditation, in which the left orientation area (on your right) is markedly decreased compared to the right side. (The darker the area, the more activity; the lighter the area, the less activity.) The images are presented in a gray scale because this provides more contrast on the printed page. The images on a computer screen, however, are usually displayed in color.

about what Robert's brain was doing during the peak moments of his meditation.

We aren't disappointed. The finished scan images show unusual activity in a small lump of gray matter nestled in the top rear section of the brain (see Figure 1.1). The proper name of this highly specialized bundle of neurons is the posterior superior parietal lobe, but for the purposes of this book, Gene and I have dubbed it the orientation association area, or OAA.[9]

The primary job of the OAA is to orient the individual in physical space—it keeps track of which end is up, helps us judge angles and distances, and allows us to negotiate safely the dangerous physical landscape around us.[10] To perform this crucial function, it must first generate a clear, consistent cognition of the physical limits of the self. In simple terms, it must draw a sharp distinction between the individual and everything else, to sort out the you from the infinite not-you that makes up the rest of the universe.

It may seem strange that the brain requires a specialized mechanism to keep tabs on this you/not-you dichotomy; from the vantage point of normal consciousness, the distinction seems ridiculously clear. But that's only because the OAA does its job so seamlessly and so well. In fact, people who suffer injuries to the orientation area have great difficulty maneuvering in physical space. When they approach their beds, for example, their brains are so baffled by the constantly shifting calculus of angles, depths, and distances that the simple task of lying down becomes an impossible challenge. Without the orientation area's help in keeping track of the body's shifting coordinates, they cannot locate themselves in space mentally or physically, so they miss the bed entirely and fall to the floor; or they manage to get their body onto the mattress, but when they try to recline they can only huddle awkwardly against the wall.

In normal circumstances, however, the OAA helps create such a distinct, accurate sense of our physical orientation to the world that we hardly need to give the matter any thought at all. To do its job so well, the orientation area depends on a constant stream of nerve impulses from each of the body's senses. The OAA sorts and processes these impulses virtually instantaneously during every moment of our lives. It manages a staggering workload at capacities and speeds that would stress the circuits of a dozen super computers.

So, not surprisingly, the baseline SPECT scans of Robert's brain taken before his meditation, while he was in a normal state of mind, show many areas of Robert's brain, including the orientation area, to be centers of furious neurological activity. This activity appears on the scans in vibrant bursts of brilliant reds and yellows.

The scans taken at the peak of Robert's meditative state, however, show the orientation area to be bathed in dark blotches of cool greens and blues—colors that indicate a sharp reduction in activity levels.

This finding intrigued us. We know that the orientation area never rests, so what could account for this unusual drop in activity levels in this small section of the brain?

As we pondered the question, a fascinating possibility emerged: What if the orientation area was working as hard as ever, but the incoming flow of sensory information had somehow been blocked?[11] That would explain the drop in brain activity in the region. More compellingly, it would also mean that the OAA had been temporarily "blinded," deprived of the information it needed to do its job properly.

What would happen if the OAA had no information upon which to work? we wondered. Would it continue to search for the limits of the self? With no information flowing in from the senses, the OAA wouldn't be able to find any boundaries. What would the brain make of that? Would the orientation area interpret its failure to find the borderline between the self and the outside world to mean that such a distinction doesn't exist? In that case, the brain would have no choice but to perceive that the self is endless and intimately interwoven with everyone and everything the mind senses. And this perception would feel utterly and unquestionably real.

This is exactly how Robert and generations of Eastern mystics before him have described their peak meditative, spiritual, and mystical moments. In the words of the Hindu Upanishads

> As the river flowing east and west
> Merge in the sea and become one with it,

Forgetting that they were ever separate rivers,
So do all creatures lose their separateness
When they merge at last into[12]

Robert was one of eight Tibetan meditators who participated in our imaging study. Each was subjected to the same routine, and in virtually every case, the SPECT scans showed a similar slowing of activity in the orientation area, occurring during the peak moments of meditation.[13]

Later, we broadened the experiment and used the same techniques to study several Franciscan nuns at prayer.[14] Again, the SPECT scans revealed similar changes that occurred during the sisters' most intensely religious moments. Unlike the Buddhists, however, the sisters tended to describe this moment as a tangible sense of the closeness of God and a mingling with Him.[15] Their accounts echoed those of Christian mystries of the past, including that of thirteenth-century Franciscan sister Angela of Foligno: "How great is the mercy of the one who realized this union . . . I possessed God so fully that I was no longer in my previous customary state but was led to find a peace in which I was united with God and was content with everything."

As our study continued, and the data flowed in, Gene and I suspected that we'd uncovered solid evidence that the mystical experiences of our subjects—the altered states of mind they described as the absorption of the self into something larger—were not the result of emotional mistakes or simple wishful thinking, but were associated instead with a series of observable neurological events, which, while unusual, are not outside the range of normal brain function. In other words, mystical experience is biologically, observably, and scientifically real.

This result did not surprise us. In fact, it was exactly what all our previous research had predicted. For years, we had scoured the scientific literature for studies examining the relationship between religious practices and the brain, searching for insights into the biology of belief. Our approach had been broad and inclusive. We found some studies that examined simple physiology—for example, they measured changes in blood pressure of people as they meditated. Other studies aimed at loftier stuff such as measuring the healing powers of prayer. We read the research on near-death experiences, studied mystical states induced by epilepsy and schizophrenia, looked at the data on hallucinations triggered by drugs as well as electrical stimulation of the brain.

Besides our scientific readings, we also looked for descriptions of the mystical components of world religions and mythologies. Gene, in particular, researched the ritual practices of ancient cultures and looked for a relationship between the emergence of ritual behavior and the evolution of the human brain. An abundance of information exists that is relevant to the relationship between religious ritual and the brain, but little of it had been sorted or synthesized into a coherent framework.

Yet as Gene and I sifted through mountains of data on religious experience, ritual, and brain science, important pieces of the puzzle came together and meaningful patterns emerged. Gradually, we shaped a hypothesis that suggests that spiritual experience, at its very root, is intimately interwoven with human biology. That biology, in some way, compels the spiritual urge.

The SPECT scans allowed us to begin testing this hypothesis by observing the actual brain activity of people engaged in spiritual practices. The images do not prove our hypothesis beyond doubt but they strongly support it by showing that, in moments of spiritual behavior, the human brain appears to behave as our theory predicts that it would.[16] These encouraging results deepened our enthusiasm for our work, and sharpened our interest in the fascinating questions provoked by our years of research. Questions like: Are human beings biologically compelled to make myths?

What is the neurological secret behind the power of ritual? Are the transcendent visions and insights of the great religious mystics based on mental or emotional delusions, or are they the result of coherent sensory perceptions shaped by the proper neurological functioning of sound, healthy minds? Could evolutionary factors such as sexuality and mating have influenced the biological development of religious ecstasy?

As we labored to better understand the implications of our theory, we found ourselves confronted again and again by one question that resonated more deeply than any other: Had we found the common biological root of all religious experiences? And if so, what did such an understanding say about the nature of the spiritual urge?

A skeptic might suggest that a biological origin to all spiritual longings and experiences, including the universal human yearning to connect with something divine, could be explained as a delusion caused by the chemical misfirings of a bundle of nerve cells.

But the SPECT scans suggested another possibility. The orientation area was working unusually but not improperly, and we believe that we were seeing colorful evidence on the SPECT's computer screen of the brain's capacity to make spiritual experience real. After years of scientific study, and careful consideration of our results, Gene and I further believe that we saw evidence of a neurological process that has evolved to allow us humans to transcend material existence and acknowledge and connect with a deeper, more spiritual part of ourselves perceived of as an absolute, universal reality that connects us to all that is.

The goal of this book is to present the surprising context for these hypotheses. We will examine the biological drive that compels us to make myths, and the neurological machinery that gives those myths shape and power. We'll discuss the connection between myth and ritual, and show how the neurological effects of ritual behaviors create those brain states associated with a range of transcendent experiences, from the mild sense of spiritual communality felt by the members of a congregation, to the deeper states of unity triggered by more intense and prolonged religious rites. We'll show that the profound spiritual experiences described by saints and mystics of every religion, and in every period of time, can also be attributed to the brain's activity that gives ritual its transcendent powers. We'll also show how the mind's need to understand these experiences can provide a biological origin for specific religious beliefs.

Sadly, my partner and friend Gene d'Aquili died shortly before work on this book began, and his contributions are sorely missed. It was Gene who first encouraged my interest in studying the link between the mind and the spirit, and who taught me to look with new eyes upon the convoluted landscape of the miraculous organ that hides inside our skulls. The work we did together—the scientific research that lay the groundwork for this book—has forced me time and again to reconsider my basic attitudes about religion and, for that matter, my attitudes about life, reality, and even my sense of self. It has been a transforming journey, a journey of self-discovery toward which I believe our brain compels us. What lies ahead in these pages is a journey into the deepest mysteries of the mind, and to the very center of the self. It begins by asking the simple question: How does the brain tell us what is real?

2

BRAIN MACHINERY

The Science of Perception

In the early 1980s, scientists at a leading university robotics center watched as their newly created robot trundled from one end of its universe to the other in jittery fits and starts. That "universe" was no more than twenty feet across, consisting entirely of a cluttered basement storage room in a university building. Yet this was the only world that the robot's programming enabled it to know.

The robot itself—an ungainly stack of computer processing units bolted to a rolling metal frame—was a rather homely traveler. It had been built for smarts, not beauty, and its powerful computer brain had been loaded with specially written software designed to help it "think" its way across the room. The robot had also been equipped with a rudimentary sense of vision, supplied by a video camera bolted to its metal frame, which would feed its digital brain with the "sensory" input needed to complete the trip successfully.

The robot's objectives, as assigned by the scientists, were simple: to use its robotic vision to navigate safely through the crowded room, find the door leading to the hallway, and swing the big door open. The casual observer might find this an unworthy challenge for such an advanced machine, but the scientists knew the mission would stretch the robot's computational powers to the limit. They hoped the experiment would tell them important things about the ability of artificial intelligence systems to perceive and interact with their surroundings while on the move.

In this experiment, movement was the issue, and the demands it placed upon the robot's computerized brain were apparent in the intense effort the robot poured into every painstaking step: even the slightest forward motion was preceded by an agonizingly prolonged analytical pause—the challenge of sidestepping a desk, for example, might delay the robot for hours.

To understand why the robot's progress was so excruciatingly slow, we have to understand the primitive fashion in which it perceived its surroundings. The robot's only source of information about the landscape that it was trying to traverse came in the form of the visual images fed to its computers by the video camera bolted to its frame. The robot depended upon these visual images to orient itself to the surrounding world, but each time the robot inched forward those images subtly changed: angles and distances shifted, shadows seemed to drift, some objects seemed closer, while others seemed farther away.

The room was not physically changing, of course; what changed was the robot's physical relationship to each item in the room. Each time the robot moved forward, it saw a different picture of its world. These pictures were different in very subtle degrees, but those subtle differences were enough to bring the brainy machine to a standstill. Its processors lacked the computational power, and its software lacked the complexity, to understand that the world portrayed in the later sets of images was nothing more than a slightly altered version of the world it had seen before.

As far as the robot was concerned, *any* change meant *total* change, and each new image was showing him a completely new and different universe. Experience from the "old" world did not carry over into the new one—reality, for the robot, did not flow continuously from one moment to the next—so each new visual

image represented an entirely new reality that had to be fathomed from scratch.

This processing placed a staggering load upon the robot's digital brain, and as a result, its forward progress was torturously slow. Finally, some ten hours after it embarked, it made its way to the door that was its destination, grasped the metal lever in a crude robotic claw, and slowly tugged the door open.

The completion of the journey triggered a short celebration among lab staffers. Afterward, the robot was rolled back to its starting point and instructed to repeat the trip. Dutifully, the robot began another painfully slow trudge across the storeroom, and after many hours of hard work, it once again stood at its destination. But as its cameras scanned the door, and its computer brain compared the resulting images to the visual template stored in its memory circuits, the robot's progress ground to a shuddering halt. Someone had pressed short strips of plastic tape onto the door to form a small black *X*. The *X* changed everything. The robot knew nothing of doors with *X*s. Nothing in its silicon-based sensibilities hinted that a door might be marked with an *X* and still be considered a door. Because of the *X*, the door's "doorness," for the robot, had dissolved, and the robot saw no choice but to turn away and continue its search elsewhere.

The experiment just described happened nearly twenty years ago, when the age of high technology was first gathering steam, and the promise of artificial intelligence was just beginning to be enthusiastically explored. Since then, generations of new computers have been created, each with successively larger memory banks, faster processing speeds, and more impressive number-crunching capabilities. Marvels such as voice recognition and virtual reality have become commonplace, and our fastest computers can solve more

mathematical equations in a heartbeat than Newton could have worked out in five lifetimes.

Despite two decades of dramatic technological advances, however, even the fastest and most powerful artificial intelligence systems still lack the ability to create as smooth and fluid a rendition of reality as the brain. And when these machines are asked to process information in a way that would allow a robot to move purposefully across a room, even the best of them are easily outperformed by a toddler, a cat, or a hamster.

These shortcomings of artificial intelligence systems are, of course, no reflection upon the brilliance of scientists working to create smart machines. They do, however, illustrate how difficult it is to weave billions of bits of disembodied data into a uniform, dynamic, durable rendition of a "world" within which an individual organism can safely and productively move about. Yet this is exactly the feat that even the humblest living creatures must constantly accomplish. It is, in fact, a fundamental requirement of survival: Organisms must tirelessly process a torrent of constantly shifting sensory data. They must sort it, process it, weave it into some useful rendition of reality, and then move about freely within that reality in ways that best enhance their chances of survival.

In basic terms, an animal's survival depends upon its ability to negotiate its environment in order to have the best chance of finding mates and food, while keeping low its chances of falling off a cliff or blundering into the path of a hungry predator. The easily disoriented robot mentioned above would fail spectacularly at all these tasks. If it were an edible living thing, it would be easily preyed upon. Monkeys, rabbits, mice, and indeed any other mobile creatures that can swiftly perceive and react to the ever-shifting dangers and opportunities that fill the world around them would survive longer.

The most likely reason that living things are capable of such re-

markable sensory processing, while cutting-edge computers are not, is that their intricate neural networks, which interpret sensory input, were not logically engineered in top-down fashion by scientists. These organic, internal networks were assembled, neuron by neuron and from the bottom up, over millions of years of evolutionary trial and error.[1] Shaped by the immediate and specific problems of survival, over countless generations of genetic fine-tuning, these neural networks developed to levels of complexity and elegant integration that even the most brilliant software engineers can only dream of approaching. The goal of every living brain, no matter what its level of neurological sophistication, from the tiny knots of nerve cells that govern insect behavior on up to the intricate complexity of the human neocortex, has been to enhance the organism's chances of survival by reacting to raw sensory data and translating it into a negotiable rendition of a world.

In all living things, the basic unit upon which brains rely to accomplish their function—the nerve cell—is very similar.[2] The neural systems of even the most primitive creatures operate according to the same basic principles of chemical stimulation and electrical conduction that propel human neurobiology. A simple flat worm,[3] for example, may have only a few hundred nerve cells in its entire anatomy, but the process by which this rudimentary neural network guides the worm's simple repertoire of behaviors—nourishing itself, reproducing, and avoiding potential danger—is the same process that, when elaborated and multiplied into the staggering intricacy of the human brain, powered Einstein's legendary thought experiments and created the poetry of Shakespeare.

The expansive neurological distance between the human brain and the nervous system of a worm is difficult to measure, but it is not infinite. The difference is primarily a matter of complexity. Neurologically speaking, in fact, complexity is primarily what separates the worm from the toad, the toad from the chimp, and the chimp from, say, Stephen Hawking.

The evolution of animal brains is generally marked by an increase in complexity.[4] The result of this complexity has been to provide organisms with the ability to perceive their environment with ever-increasing precision, and to react to that environment with a more versatile, more effective range of adaptive responses. In primitive organisms like the worm, nervous systems are comprised of a simple string of nerve cells, which provide only the coarsest interpretation of reality, allowing only crude approach/avoid reactions.

As you trace the evolution of species, however, those neural strings become longer and more intricate. They begin to loop and tangle. Elaborate neural networks evolve, and as nerve cells increase in number, first by millions, then by billions, they begin to cluster into highly specialized structures that allow ever more sophisticated processing of sensory information. Eventually, connecting circuits develop among these structures, which allow them to share and integrate information to produce rich, multilayered perceptions of the environment and highly efficient ways of adapting to it.

The billowing complexity that characterizes the evolution of neurological systems reaches its fullest point so far in the elegant engineering of the human brain. By virtue of its highly developed neural architecture, the brain provides human beings with a multisensory and multiperceptual understanding of the world around them. It also enables a vast repertoire of sophisticated behaviors with which to react to the threats and opportunities that the environment presents.

Humans have the capacity to anticipate good and bad situations, imagine alternatives and assorted potential outcomes, and make plans to try to ensure an optimal result. Thanks to their big, complex brains, early humans learned to store food for the future, plant crops, and dig wells. To better their chances of survival, they banded together into tribes and clans and developed ways of communicating, which allowed them to hunt, share resources, and defend themselves more efficiently. As their societies evolved, humans found more and more sophisticated ways to gain control over their environment in the form of cities, nations, governments, religions, culture, technologies, and eventually science.

The brain's functions that have allowed humans to accomplish these achievements are variously described as: creativity, genius, insight, and inspiration. But none of these quintessentially human accomplishments would have been possible without the brain's ability to generate rich, effective, and meaningful perceptions of the world.

The average human brain weighs about three and a half pounds. It's roughly the size of a large head of cauliflower, and resembles, in color and consistency, a generous blob of extra-firm tofu. Small ligaments help moor the brain to the walls of the bony skull, and a thin layer of fluid provides a cushion between the skull and the brain's convoluted outer surface. Those distinctive convolutions are the clustered contours of the various individual structures that make up the brain's conglomerate structure. Each of these structures has a set of highly specialized functions, but each also cooperates with the rest of the brain as a whole in complex and elegant ways, giving it the ability to channel, interpret, and respond to the rush of information flooding the body's neural pathways.

There are two basic ways to determine the function of any given brain structure. The first involves the study of brains damaged in some way, most often, by a tumor, trauma, or stroke. By correlating the damaged area with the corresponding loss of function, for example, scientists have learned that damage to the occipital lobe results in the impairment of vision, and that damage to the temporal lobe can affect the ability to speak.

A second way of determining the functions of the brain is through the study of brain images obtained while subjects perform specific behaviors or tasks. These are called activation studies and show which areas of the brain are activated when a certain behavior is performed.

It is beyond the scope and purpose of this book to provide an exhaustive anatomical study of the brain. In order to understand the relationship between the brain and spirituality, however, it's important to have an understanding of basic brain functions. With that purpose in mind, we have focused on the individual brain structures we believe are most pertinent to the phenomenon of spiritual experience. Some of these structures are involved in the generation of emotional and neurobiological states and will be discussed in the following chapter. In this chapter, we will focus on the cerebral cortex, often considered the seat of our human nature. The cerebral cortex performs most of the brain's higher cognitive functions, and its various processing centers, or association areas, assemble the streams of neural impulses into the meaningful perceptions with which the brain makes sense of the world.

In all our discussions, we have made an effort to use terms and descriptions that will be accessible to a nonprofessional audience. These will not necessarily be the terms that neuroscientists would use. Furthermore, some of our discussions are based upon empirical fact and others upon hypothesis. We will strive to make a clear distinction between the two. All of our comments, however, whether factual or theoretical, are based on solid scientific research. Interested readers will find references to this background information in the notes section at the end of the book.

WHAT MAKES US HUMAN: THE CEREBRAL CORTEX

Most of the human brain is contained in the familiar folds of the cerebral cortex, where all high-order cognitive functions occur.

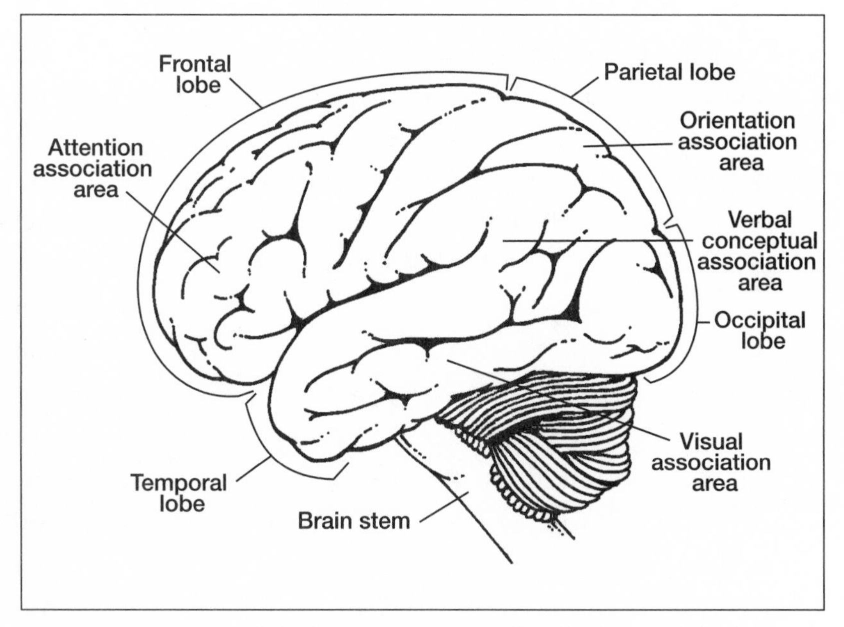

Figure 2.1: A view of the brain from the side (the front of the brain is to the left).

The vast majority of the cerebral cortex is referred to as the neocortex, because it is the most recently evolved region of the brain. The evolution of this "new cortex" gave us the cerebral intelligence that separates humans so decisively from other animals and enables us to create language, art, myth, and culture.

The cortex is connected to the body by more primitive "subcortical" structures that regulate basic life-support systems, the activity of hormones, and primal emotions. The subcortical structures connect the neocortex to the brain stem, which in turn connects the brain to the spinal cord and the biological processes of the body (see Figure 2.1).[5] So the cerebral cortex is also an important center of sensory and motor control. It is where mind and body come together and create our self-image and our view of the world.

The cerebral cortex is divided into the left and right hemispheres, and each hemisphere is further divided into four large

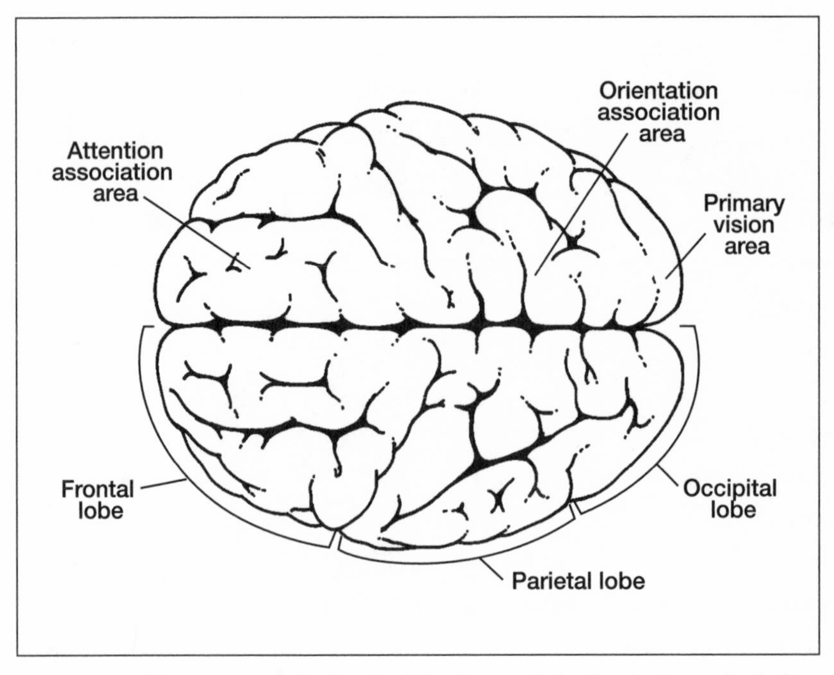

Figure 2.2: Top view of the brain (the front of the brain is to the left).

structures known as lobes (see Figures 2.1 and 2.2). The temporal lobe, located along the sides of the head, is associated with language and conceptual thinking; the occipital lobe, at the back of the head, with vision; the parietal lobe, which sits beneath the crown of the skull is home to sensory perception, visual-spatial tasks, and body orientation; and the frontal lobe, situated directly behind the forehead, is associated with attention and initiating muscle activity.

The two hemispheres of the brain are similar in appearance and, to some extent, function. For example, the left hemisphere receives and analyzes sensations from the right side of the body and governs right-side motor activity, while the right hemisphere does the same for the body's left side. Both hemispheres also contain centers for processing language, which, when working in concert, give us the power of expressive verbal communication.

At the same time, however, there are important differences in the way the hemispheres work. The classic understanding is that the left hemisphere is more analytically inclined and is recognized as the center of verbal language and mathematical processes.[6] The right hemisphere works in a more abstract, holistic way, as a center of nonverbal thought, visual-spatial perceptions, and the perception, modulation, and expression of emotions. But we want to emphasize that both hemispheres can contribute to similar mental functions.

Because the left cerebral hemisphere generally provides the neurological basis for oral and written language, and because those functions are so important in the formulation and expression of conscious thought, the left is often referred to as the "dominant" hemisphere. But the proper functioning of the brain typically requires the coordinated interaction of both sides of the cortex. This communication between the hemispheres is made possible by networks of connecting nerve fibers. Studies have shown that due to the nature of their neural wiring, these connecting structures cannot convey complex thoughts and perceptions, only nuances.

For instance, if you are trying to solve a geometrical puzzle, such as how to fit a square into a circle, the analytical thought process is mostly happening on the left side. There, the brain is working with the logic of measuring lengths and circumference, while the right side is looking at the shapes as wholes. Because the connectors between right and left are not conveying detailed input, the right brain is not fully informed of the analyses of the left brain. The right side does not fully participate in the solution with its images. However, when the imaged solution on the right side of the brain matches the analyses on the left, both hemispheres declare the problem solved through electrical impulses to the emotional center of the brain.[7]

As these simplistic representations of mental activity of thought or perception travel from one side of the brain to the other, they profoundly affect the brain's accuracy in experiencing the world.[8]

This point can be further illustrated by examining the way in which the brain turns raw auditory impulses into meaningful verbal understanding. The process begins as impulses from primary hearing centers reach the brain's main language center, usually located in the left cerebral hemisphere. For instance, the brain hears the sound "right," which sounds exactly the same as "write" and "rite," and this auditory input is converted into intelligible words and sentences and understood logically, in the context of grammar and syntax.

Meanwhile, a secondary language center, located in the right hemisphere, is informed of the left-side activity by impulses traveling across the connecting structures as well as directly from the primary hearing centers. Immediately, the right side, calling upon its powers of abstract, intuitive perception, begins to interpret the emotional tones and verbal inflections that give spoken language its subtle shades of meaning.

The importance of this operation becomes clear when the language area in the right side is destroyed. In such a case, the language center in the left hemisphere can still logically comprehend the literal meaning of words and phrases, but without the right brain's more intuitive contributions, the emotional intent of words would be lost.[9] As a result, a person in such a state might hear the phrase "Get out of here!" and have no way of judging the emotional tone of the remark: Is it a hostile command or a joking expression of disbelief?

The importance of left-brain/right-brain cooperation is also demonstrated when the structures connecting them are surgically severed, as they sometimes are in the treatment of epilepsy, to prevent localized seizures from affecting the entire brain. Research shows that in such split-brain cases, the brain generates what seems to be two separate consciousnesses. Research on split-brain patients led brain scientist and Nobel laureate Roger Sperry to conclude, "Everything we have seen indicates that the surgery has left these people with two separate minds, that is, two separate spheres of consciousness. What is experienced in the right hemisphere seems to lie entirely outside the realm of the left hemisphere."[10]

For example, if a split-brain patient is shown a picture of a hammer in a way that allows visual input to reach only the left side of the brain, the patient is able to state verbally that he has seen a hammer. If, however, the patient is shown the picture in a way that visual impulses reach only the right side of the brain, the patient can't verbally describe what he's seen.[11] Lacking the left side's ability to turn visual input into logically verbalized concepts, the right side has no way to convert the image into words. But if asked to draw a picture of what he has been shown, the patient is able to sketch the likeness of a hammer.

Interestingly, if the image shown exclusively to the right hemisphere creates a strong emotional response, the left side will search for a logical reason for those emotions. A split-brain patient shown a photograph of Hitler only in the right hemisphere, for example, might exhibit facial expressions indicating anger or disgust. But when asked to explain those emotions, the patient will often invent an answer, such as "I was thinking about a time when someone made me angry."

Based on such research, investigators have concluded that both hemispheres of the brain are capable of some kind of awareness, but their methods of experiencing and expressing it are very different. Clearly, the generation of human conscious awareness, in all its multilayered fullness, depends upon the harmonious integration of both sides of the brain.

The highly complex structures contained in the cerebral cortex give us the distinctive quality of mind that we think of as characteristically human, but cortical structures are also involved in gathering together streams of sensory input and assembling them into the vivid sensory perceptions that give the mind its content.

ASSEMBLING PERCEPTIONS

The basic functional unit of the human nervous system is the neuron,[12] the tiny, spindly cell that, when arranged into intricately woven chains of long neural pathways, carries sensory impulses to the brain. At the basic level, sensory data enters the neural system in the form of billions of tiny bursts of electrochemical energy gathered by countless sensors in the skin, eyes, ears, mouth, and nose. These neural impulses race along neural pathways, cascading like a line of falling dominoes, leaping synaptic gaps and triggering the release of chemical neurotransmitters as they carry their sensory messages toward the brain.

Once inside the brain proper, sensory information is channeled along the appropriate neural pathways—optical input travels the pathways of the brain's visual system, for example, while impulses from the sense of smell are channeled along olfactory circuits. As they flash through the brain, individual impulses are routed and rerouted to appropriate processing areas.[13] Here they are sorted, cross-referenced, amplified or inhibited, integrated with input from emotional centers and other senses, and finally assembled into a perception that has a useful, individual meaning to the owner of that particular brain.

The first level of sensory processing occurs in the *primary receptive areas* dedicated to each of the five sensory systems. These areas receive unprocessed input directly from the senses and assemble that raw data into rough, preliminary perceptions. These perceptions then travel to the *secondary receptive areas*, each of which is also dedicated to a specific sensory system, where they are further refined.

Sensory perceptions then move to the *association areas*, where the most sophisticated processing occurs. These structures are called association areas because they gather together, or "associate," neural information from various parts of the brain. At this highest level, information from a single sense is integrated with information from all the other senses to create the rich, multidimensional perceptions that form the building blocks of consciousness. The association areas eventually tap into memory and emotional centers to allow us to organize and respond to the exterior world in the most complete way possible.

To understand how the brain turns raw information into finished, useful perceptions, let's follow the path of sensory information flowing through the visual processing system.

A visual image originates in the electrochemical impulses streaming into the brain along the optic nerve. The first stop for these impulses once they arrive in the cortex is the primary visual area, where they are translated into crude visual elements—a jumble of abstract lines, shapes, and colors.

The rough visual patterns discerned by the primary visual area can't be perceived by the conscious mind, but there is evidence that the brain can become aware of them on an unconscious level. The phenomenon of "blindsight" is an interesting example.[14] This condition occurs when serious damage to the primary visual association area prevents visual input from reaching the secondary level, where, under normal circumstances, it would eventually be refined into the finished images that eventually enter consciousness. People who suffer such damage to the primary area consider themselves totally blind; yet despite their lack of conscious sight, they are able to reach out to objects in front of them, correctly answer questions about objects they are "looking" at, and may even make their way through crowded rooms. Their "blindsight" is the result of the brain's unconscious ability to recognize unformed visual data that, as raw as it may be, is apparently informative enough to allow the individuals to safely negotiate their physical surroundings.

There is no definitive explanation for the phenomenon of blindsight. It could be that alternate visual pathways develop to bypass the primary processing areas and carry impulses to the secondary level. It's also possible that small portions of the damaged primary area continue to function. Whatever the precise mechanism of blind-sight may be, the condition gives us a fascinating insight into the brain's processing systems.

In a normally functioning brain, the abstract shapes and colors processed in the primary visual area would be further organized in the secondary area, and a recognizable image would begin to form. If, for example, the sensory input in question is the result of gazing at a poodle, then the patterns formed at the primary stage will be assembled, or associated, into a composite shape resembling a small, curly-haired dog.

This finished image may be accessible to the conscious mind; however, because the brain has not yet combined the image with the components of memory and emotion, which will allow the image to be associated with the concept of "dog," it lacks context and meaning, and will be just a free-floating picture until the final stage of processing can be performed.

That final stage is carried out in the visual association area, where the image of the poodle is associated with input from other parts of the brain, bestowing it with dimensional fullness and emotional meaning. Olfactory areas, for example, alert the mind to a strong, familiar scent. The auditory areas contribute the sound of playful barking. Input from the memory areas place the image in the context of your past experience, and finally, the emotional areas trigger a sudden surge of affection, because the image is no longer just an image, but has developed into the realistic, fully integrated experience of gazing upon your own beloved dog. Someone who has had a negative or hurtful encounter with a dog, however, or perhaps has read alarming news about dogs that attack people, will see your dog as a potential danger and threat.

The visual association area is vital to the meaningful interpretation of the brain's perceptions. Damage to this area often disrupts an individual's ability to recognize friends, family members, and pets. It does not disrupt his ability to see these things, but only to place them in a meaningful context of emotion and memory. In some cases, victims of such damage are unable even to recognize their own faces in the mirror.

The visual association area may also play a prominent role in religious and spiritual experiences that involve visual imagery. For example, the visual association area is likely to be active in individuals who use images (such as of a candle or a cross) to help facilitate meditation or prayer. Furthermore, spontaneous visions that occur during meditation and prayer or that are associated with unusual spiritual states such as near-death experiences may also originate in this area. We know this, in part, because electrical stimulation of this area results in various types of visual experiences.[15] In addition, this area is closely tied to the brain's memory banks, so it is possible that stored visions are remembered or associated—perhaps not fully consciously—with later religious and spiritual experiences.

UNDERSTANDING AND RESPONDING TO THE WORLD AROUND US

There are several association areas in the cerebral cortex designed to put all of this sensory information together. Some are dedicated to a single sense, others receive input from more than one sensory system. All of them process and integrate information with the same ultimate goal: to enrich our understanding of the world outside the skull by identifying specific objects and determining what our emotional, cognitive, and behavioral responses to them should be.

We believe that four association areas in particular play an important role in producing the mind's mystical potential. The visual association area, which we've just discussed, is one. The others are the orientation association area, the attention association area, and the verbal conceptual association area.[16] A brief description of each of those areas follows, along with some evidence of the roles they may play in the mystical capabilities of the mind.

Defining the Self: The Orientation Association Area

The orientation association area, situated at the posterior section of the parietal lobe, receives sensory input from the sense of touch as well as from other sensory modalities, especially vision and hearing. These give it the ability to create a three-dimensional sense of "body" and to orient that body in space.[17]

There are two orientation areas, one located in each hemisphere of the brain that performs related but distinct operations, as shown in brain imaging studies. The left orientation area is responsible for creating the mental sensation of a limited, physically defined body, while the right orientation area is associated with generating the sense of spatial coordinates that provides the matrix in which the body can be oriented. In simpler terms, the left orientation area creates the brain's spatial sense of self, while the right side creates the physical space in which that self can exist.

The process through which the brain might construct these fundamental categories of self and not-self are not clearly understood, though researchers have found some tantalizing clues. For example, we know that certain neurons in the left orientation area respond only to objects within arm's reach, while others respond only to objects just beyond. This fascinating finding has led some researchers to postulate that the distinction between self and other may have its roots in the ability of the left orientation area to judge between these two simple categories of reality—that which can be grasped and that which can't.[18]

Whatever the origins of the orientation response might be, and however the brain supports it, the important point is that by working in concert, the two sides of the orientation association area are able to weave raw sensory data into the vivid, complex perception of a self and into a world in which that self can move. The fact that this "self" is a mental representation, and that it is assembled from bits of raw sensory data, does not mean, of course, that the physical body or the world around it does not exist. The point is that the only way the mind can know the self, and experience the difference between the self and the rest of reality, is through the elaborate, restless efforts of the brain.

We believe that the orientation association area is extremely important in the brain's sense of mystical and religious experiences, which often involve altered perceptions of space and time, self and ego. Since the orientation association area is instrumental in shaping these basic perceptions, it must somehow be an integral part of spiritual experience.

The Seat of the Will: The Attention Association Area

The attention association area, also known as the prefrontal cortex of the brain, plays a major role in governing complex, integrated bodily movements and behaviors associated with attaining goals.[19] For instance, this area helps the body organize the behaviors necessary for reaching desired objects or moving toward some chosen destination.[20] On an even more complex level, the attention area seems to be critically involved in organizing all goal-oriented behaviors and actions, even purposefully directed patterns of thought that are intended to focus the mind upon a particular object or idea.

This structure is so heavily involved in such intentional behavior, in fact, that a number of researchers think of the attention area as the neurological seat of the will.[21] Several studies suggest that the attention area is able to focus the mind upon important tasks through a process neurologists describe as "redundancy."[22] Redundancy allows the brain to screen out superfluous sensory input and concentrate upon a goal. It's what allows you to read a book in a noisy restaurant or daydream while walking along a crowded street.

The ability of the attention area to form intentions and act upon them is supported by research which shows that damage to this area of the brain results in a loss of the ability to concentrate, plan future behavior, and carry out complex perceptual tasks that require sharp mental focus or sustained attention. Victims of such damage, for example, are often unable to complete long sentences or plan a schedule for the day. They also frequently exhibit emotional flatness, a lack of will, and a profound indifference to events in the environment.[23] These findings, as well as brain imaging studies, indicate that the frontal lobes are involved in the processing and control of emotion in association with the limbic system, with which it has multiple interconnections.

To illustrate how the attention association area works, one experiment showed that subjects who counted out loud had increased brain activity primarily in the motor area, corresponding to the movement of the tongue, lips, and mouth. Subjects who counted to themselves, however, showed increased activity in the attention association area, indicating this area's involvement in focusing the mind especially when no motor activity is involved.[24]

The attention association area has already been shown to be important in various religious and spiritual states. Brain imaging studies such as ours and that of several other investigators have found increased activity in the attention association area during certain types of meditation.[25] A number of other studies have shown that during sustained attention the electrical activity in the frontal lobe of the brain changes as measured by electroencephalography (EEG) and that this change is particularly pronounced among Zen practitioners during meditation.[26]

Although there is a great deal of data from EEG readings of people concentrating intensely, there is unfortunately only one report of an EEG reading taken when a subject experienced a near "peak" experience. Because peak experiences are quite rare, the likelihood of catching one when the subject is hooked up for electrophysiological readings is slim. The EEG record of this subject during meditation demonstrated significant EEG changes particularly in the attention association area as well as in the orientation association area.

We believe that part of the reason the attention association area is activated during spiritual practices such as meditation is because it is heavily involved in emotional responses—and religious experiences are usually highly emotional. So it seems reasonable that the attention association area must have some important interaction with other brain structures underlying emotion during meditative and religious states.

Naming and Cataloging the World: The Verbal Conceptual Association Area

The verbal conceptual association area, located at the junction of the temporal, parietal, and occipital lobes, is primarily responsible for generating abstract concepts and for relating those concepts to words.[27] Most of the cognitive operations required in the using and comprehending of language—the comparison of concepts, the ordering of opposites, the naming of objects and categories of objects, and the high-order grammatical and logical functions—are carried out in the verbal conceptual association area. These operations are crucial to the development of consciousness and the expression of consciousness through language.

The verbal conceptual association area is extremely important for all of our mental functioning, and it should come as no surprise that it is equally important in religious experience, since almost all religious experiences have a cognitive or conceptual component—that is, some part that we can think about and understand. A study by V. S. Ramachandran at UCLA showed that patients with temporal lobe epilepsy have a heightened response to religious language, specifically religious terms and icons. The suggestion from these findings is that the temporal lobe is very important in these experiences.[28] Furthermore, this area houses other important brain functions, such as causal thinking, that are associated with how we create myth and ultimately how myth is expressed in rituals.

These four association areas are the most complex neurological structures in the brain. Their rich, fully integrated perceptions allow us to experience reality as a vivid, cohesive whole that flows smoothly and comprehensibly from one moment to the next. The fuller these perceptions, the greater our chances for survival, which is the ultimate goal of all the neurobiological workings of the brain.

HOW THE BRAIN MADE ITS OWN MIND

As the human brain evolved something remarkable happened: The brain, with its great perceptual powers, began to perceive its own existence, and human beings gained the ability to reflect, as if from a distance, upon the perceptions produced by their own brains. There seems to be, within the human head, an inner, personal awareness, a free-standing, observant self. We have come to think of this self, with all its emotions, sensations, and cognitions, as the phenomenon of *mind*.

Neurology cannot completely explain how such a thing can happen—how a nonmaterial mind can rise from mere biological functions; how the flesh and blood machinery of the brain can suddenly become "aware." Science and philosophy, in fact, have struggled with this question for centuries, but no definitive answers have been found, and none is clearly on the horizon.

Until now, we have used the terms "brain" and "mind" somewhat loosely. For our purposes, we'll set up a couple of simple, straightforward definitions based on the ever-growing understanding of important mental processes. These definitions in particular express the ways in which the structures of the brain operate harmoniously to turn raw sensory data into an integrated perception of the world outside the skull: the *brain* is a collection of physical structures that gather and process sensory, cognitive, and emotional data; the *mind* is the phenomenon of thoughts, memories, and emotions that arise from the perceptual processes of the brain.[29]

In simpler terms, brain makes mind. Science can demonstrate no way for the mind to occur except as a result of the neurological functioning of the brain. Without the brain's ability to process various types of input in highly sophisticated ways, the thoughts and feelings that constitute the mind would simply not exist. Conversely, the brain's irresistible drive to create the most vivid, sophisticated perceptions possible means that it cannot help but generate the thoughts and emotions that are the basic elements of mind.

Neurologically speaking, then, the mind cannot exist without the brain, and the brain cannot exist without striving to create the mind. The relationship of mind and brain is so intimately linked, in fact, that it seems most reasonable to consider the terms as two different aspects of the very same thing.

Consider, for example, that the existence of a single human thought requires the highly complex interaction of hundreds of thousands of neurons. In order to separate mind from brain, it would be necessary to think of each neuron as something distinct from its function, which is a little like trying to separate the seawater that provides the substance of an ocean wave from the energy that gives the wave its shape and motion. The existence of the wave requires both elements: without energy, the wave would fall flat; without water, the wave energy would have no expression. In the same sense, it is not possible to separate individual neurons from their functions; if it were possible, then a thought could be freed from its neurological base, and the mind could be seen as something separate from the brain, a free-floating consciousness that could be considered a "soul."

Such a separation, however, would be staggeringly difficult to perform, even with a single thought involved. When you consider the vast, integrated neurological activity of the brain as a whole, differentiating neurons from their functions becomes unimaginable. Reason leads us back to the conclusion that mind needs brain, brain creates mind, and that the two are essentially the same entity, seen from different points of view.

The inexplicable unity of the biological brain and its ethereal phenomenon of mind is the first aspect of what we have defined as the mind's mystical potential. The second characteristic, which was hinted at in our SPECT scan studies, is the ability of the mind to interpret spiritual experience as real. This ability, based on the mind's capacity to enter altered states of consciousness, and to adjust its assessment of reality neurologically, is a fundamental link between biology and religion. But before we can understand the nature of that connection, we must explore the emotional and neurological components that form the brain's foundation for this mystical mind.

3

BRAIN ARCHITECTURE

How the Brain Makes the Mind

Every time that the powers of the soul come into contact with created things, they receive the created images and likenesses from the created thing and absorb them. In this way arises the soul's knowledge of created things. Created things cannot come closer to the soul than this, and the soul can only approach created things by the voluntary reception of images. And it is through the presence of the image that the soul approaches the created world, for the image is a thing that the soul creates with her own powers. Does the soul want to know the nature of a stone—a horse—a man? She forms an image.

—Meister Eckhart, "Mystiche Schriften," quoted in *Mysticism* by Evelyn Underhill

The medieval German mystic Meister Eckhart lived hundreds of years before the science of neurology was born. Yet it seems he had intuitively grasped one of the fundamental principles of the discipline: What we think of as reality is only a rendition of reality that is created by the brain.

Our modern understanding of the brain's perceptual powers bears him out. Nothing enters consciousness whole. There is no direct, objective experience of reality. All the things the mind perceives—all thoughts, feelings, hunches, memories, insights, desires, and revelations—have been assembled piece by piece by the processing powers of the brain from the swirl of neural blips, sensory perceptions, and scattered cognitions dwelling in its structures and neural pathways.

The idea that our experience of reality—all our experiences, for that matter—are only "secondhand" depictions of what may or may not be objectively real, raises some profound questions about the most basic truths of human existence and the neurological nature of spiritual experience. For example, our experiment with Tibetan meditators and Franciscan nuns showed that the events they considered spiritual were, in fact, associated with observable neurological activity. In a reductionist sense, this could support the argument that religious experience is only imagined neurologically, that God is physically "all in your mind." But a full understanding of the way in which the brain and mind assemble and experience reality suggests a very different view.

Imagine, for instance, that you are the subject of a brain imaging study. As part of this study, you have been asked to eat a generous slice of homemade apple pie. As you enjoy the pie, the brain scans capture images of the neurological activity in the various processing areas of the brain where input from your senses is being turned into the specific neural perceptions that add up to the experience of eating the pie: olfactory areas register the delightful aroma of apples and cinnamon, visual areas perceive the sight of the golden brown crust, centers of touch perceive the complex mix of crunchy and gooey textures, and the rich, sweet, satisfying flavors are processed in the areas responsible for taste. The SPECT brain scan would show all this activity in the same way that it revealed the brain activity of the Buddhists and the nuns, as blotches of bright colors on the scanner's computer screen. In a literal sense, the experience of eating the pie is all in your mind, but that doesn't mean the pie is not real, or that it is not delicious.

Similarly, tracing spiritual experience to neurological behavior does not disprove its realness. If God does exist, for example, and if He appeared to you in some incarnation, you would have no way of experiencing His presence, except as part of a neurologically generated rendition of reality. You would need auditory processing to hear His voice, visual processing to see His face, and cognitive processing to make sense of His message. Even if He spoke to you mystically, without words, you would need cognitive functions to comprehend His meaning, and input from the brain's emotional centers to fill you with rapture and awe. Neurology makes it clear: There's no other way for God to get into your head except through the brain's neural pathways.[1]

Correspondingly, God cannot exist as a concept or as reality anyplace else but in your mind. In this sense, both spiritual experiences and experiences of a more ordinary material nature are made real to the mind in the very same way—through the processing powers of the brain and the cognitive functions of the mind. Whatever the ultimate nature of spiritual experience might be—whether it is in fact a perception of an actual spiritual reality, or merely an interpretation of sheer neurological function—all that is meaningful in human spirituality happens in the mind. In other words, the mind is mystical by default. We can't definitively say why such capabilities have evolved, but we can find traces of their neurological roots in some basic structures and functions, primarily the autonomic nervous system, the limbic system, and in the brain's complex analytical functions.

YOUR AROUSAL AND QUIESCENT SYSTEMS

The arousal and quiescent systems are the most basic part of the body's nervous system, and their fibers provide a major neurological bridge between the brain and the rest of the body. With the input of various brain structures, the autonomic nervous system is responsible for regulating fundamental functions such as heart rate, blood pressure, body temperature, and digestion. At the same time, because of its connections to higher brain structures, it also has a significant relationship with many other aspects of brain activity, including the generation of emotions and mood.

The autonomic system is composed of two branches: the sympathetic and the parasympathetic nervous systems.[2] The sympathetic system is the basis of the body's fight-or-flight response, which gives us the adrenaline boost we need to escape or defend ourselves from danger. The arousal system is also activated by positive experiences—it makes the hunter's heart beat faster as he closes in upon his prey, for instance. It is also involved with mating. In fact, any situation that involves an element of survival interest will activate the sympathetic system. Whether that situation is a threat or an opportunity, the response will be the same—a surge of intense readiness or arousal. Physiologically, this is expressed as faster heart rate, higher blood pressure, quicker breathing, and increased muscle tone. In this aroused state, the body is expending energy to allow for decisive physical action.

Because of its ability to prepare the body for action, we will consider the sympathetic system, along with its connections to the brain and the adrenal glands, as the body's arousal system.

The counterbalance to this arousal function is provided by the parasympathetic nervous system. The parasympathetic system is responsible for conserving energy and for keeping all the body's basic functions in harmonic balance. It regulates sleep, induces relaxation, promotes digestion, distributes vital nutrients throughout the body, and governs the growth of cells. Because of its ability to exert a calming, stabilizing effect upon the body, we will call the parasympathetic nervous system, along with its associated structures in the upper and lower parts of the brain, the quiescent system.

In general, the arousal and quiescent systems operate in antagonistic fashion: Increased activation of one system usually results in the decreased activity of the other.[3] This allows the body and brain to act smoothly and respond appropriately to the situation at hand. For example, when faced with a threat, the quiescent system defers to the arousal function, allowing it to expend the energy that will physiologically prepare the body for action. In the same fashion, the arousal system will defer to the quiescent system when the threat is over, freeing it to lower blood pressure, slow breathing, and, in general, conserve efficiently the body's stores of fuel and energy.

These alternating interactions generally occur during routine everyday activity.[4] There is evidence, however, of cases in which both systems function at the same time when pushed to maximal levels of activity and this has been associated with extraordinary alternative states of consciousness. These unusual, altered states can be triggered by various kinds of intense physical or mental activity, including dancing, running, or prolonged concentration.[5] These states can also be intentionally triggered by specific activities that are overtly religious in nature, such as ceremonial rituals or meditation. The similarities between these intentionally and unintentionally triggered states point to a clear link between the autonomic nervous system and the brain's potential for spiritual experience.

We suspect that the autonomic nervous system is, in fact, fundamental to religious experience. Many early studies demonstrated that practices such as tantric yoga and transcendental meditation, for example, are associated with significant changes in heart rate, blood pressure, and breathing—all of which are controlled by the autonomic nervous system.[6]

These and other meditation studies have measured changes in the electrical conductance of the skin (another autonomic function), which is dependent on the amount of sweat we make in response to different situations. Obviously, when we are aroused or under stress, we sweat more. Interestingly, the studies have not always demonstrated consistent changes. While some research suggests that both systems are active during meditation, certain types of meditation appear to have a greater effect on the arousal system and others a greater effect on the quiescent system. Of course, extrapolating meditative practices and their results to all religious experience is not a simple step. But we have identified four autonomic states that we believe contribute to understanding the broad range of altered and potentially spiritual states.[7] These states can help us explore the relationship between the autonomic nervous system and religious experience.[8]

AUTONOMIC STATES AND SPIRITUAL EXPERIENCE

1. Hyperquiescence

Hyperquiescence is a state of extraordinary relaxation. The body usually experiences it only during deep sleep, but it may also occur during certain phases of meditation. It can be evoked through slow, quiet, deliberate rituals, such as chanting or group prayer. At intense levels, the body and mind have a sense of oceanic tranquillity and bliss in which no thoughts, feelings, or bodily sensations intrude upon consciousness. Buddhists describe a similar state, reached through meditation, as "access consciousness," or *Upacara samadhi.*

2. Hyperarousal

The flip side of hyperquiescence, the hyperarousal mental state is characterized by an unblocked flow of arousal and excitation, resulting in a burgeoning sense of excitement, keen alertness, and fierce concentration to the exclusion of any extraneous feelings or thoughts. Hyperarousal can result from any kind of continuous, rhythmic motor activity, such as the rapid ritual dancing of Sufi mystics or Voudon practitioners. It has also been reported by marathon runners and long-distance swimmers. In some cases, hyperarousal will occur spontaneously in individuals engaged in critical situations—a downhill skier in an Olympic race, for example, or a fighter pilot in a skirmish—who must make instantaneous decisions based upon the processing of huge amounts of sensory information. For them, normal, conscious thoughts could be disastrous distractions.

People in hyperarousal states often feel as if they are effortlessly channeling vast quantities of energy through their consciousness, resulting in the quintessential "flow" experience.[9]

3. *Hyperquiescence with Arousal Breakthrough*

Under certain unusual conditions, the quiescent branch of the autonomic system can be driven to such intense levels of activity that the normal antagonistic reaction between the sympathetic and parasympathetic systems is overwhelmed. As a result, rather than being inhibited by quiescent activity, the arousal system instead becomes highly stimulated. This neurological "spillover" or "breakthrough" can lead to intensely altered states of consciousness.

In meditation or contemplative prayer, powerful quiescent activity can result in sensations of great bliss, but when quiescent levels reach maximum, the arousal system can simultaneously erupt, causing an exhilarating rush of energy. Someone who experiences this state while concentrating upon some object—a candle for example, or a cross—may feel as if he were being absorbed into that object. Buddhists call this state of absorption *Appana samahdi.*

4. Hyperarousal with Quiescent Breakthrough

The maximal stimulation of the arousal system can also cause a spillover effect, which causes quiescent responses to surge. The resulting trancelike state is experienced as an ecstatic rush of orgasmiclike energy. This state can be induced by intense and prolonged contemplation, during rapid ritualistic dancing, and sometimes, briefly, during sexual climax. To a large degree, the arousal-quiescent systems and their four states of interaction connect the body to the mind and vice versa. Since these states are usually intimately related to emotions, the autonomic nervous system must be closely connected to the limbic system that regulates emotions. Because the autonomic nervous system is involved with how we experience positive and negative emotions, which themselves arise in the limbic system, we should track how our emotional brain creates the emotions of our mind.

THE EMOTIONAL BRAIN: THE LIMBIC SYSTEM

The human limbic system interweaves emotional impulses with higher thoughts and perceptions to produce a broad, flexible repertoire of highly complex emotional states such as disgust, frustration, envy, surprise, and delight. These emotions, although built on primal emotions found to some extent in other animals, give human beings a more diverse articulated emotional vocabulary.

Studies have also indicated that the limbic system is integral to religious and spiritual experiences. Electical stimulation of the limbic structures in human subjects produces dreamlike hallucinations, out-of-body sensations, déjà vu, and illusions, all of which have been reported during spiritual states.[10] On the other hand, when nerve input into the limbic system is blocked, visual hallucinations may also result.[11] Because of its involvement in religious and spiritual experiences, the limbic system has sometimes been referred to as the "transmitter to God."[12] Whatever its relationship to the spiritual, before it learns to transmit activity, however, the limbic system's most fundamental purpose is to generate and modulate primal emotions such as fear, aggressiveness, and rage. Possessed by most animals with central nervous systems, the structures of the limbic system are ancient, in evolutionary terms. What separates our limbic system from those of animals and our early ancestors are the many subtleties that we have developed. Jealousy, pride, regret, embarrassment, elation—all are the products of a highly developed limbic system, especially when its function is combined with the rest of the brain. So while our early ancestors may have felt a pang of disappointment at not attending their son's rock-throwing contest, we now feel full-fledged guilt.

The primary structures of the limbic system are the hypothalamus, the amygdala, and the hippocampus. All are primitive organs, but their influence upon the human mind is considerable.

The survival advantage of the limbic system is clear: It gives animals the aggressiveness they need to find food, the fear that compels them to avoid predators and other dangerous situations, and the affiliative longing—primitive "love," if you will—that drives them to mate and care for their young. In human beings, the primal feelings produced by limbic activity are integrated with higher cognitive functions from the neocortex, resulting in richer, more varied, emotional experiences.

The Master Controller: The Hypothalamus

From an evolutionary perspective, the oldest structure in the human limbic system is the hypothalamus, which sits near the upper end of the brain stem. Even though the hypothalamus is part of the limbic system, it can be thought of as the master control for the autonomic nervous system.[13] The hypothalamus has two basic sections: the inner section, which is connected to the quiescent system and can generate calming emotions, and the outer edge, which is an extension of the arousal system into the brain. The hypothalamus can help create basic emotions such as rage and terror as well as positive states ranging from moderate pleasure to bliss.

One of the major roles of the hypothalamus is to link the operations of the autonomic system to the higher structures of the brain's neocortex. It provides the key link through which the brain can instruct the autonomic system to regulate body functions. It is also the gateway through which autonomic impulses can be relayed to higher structures in the brain for processing and understanding. Thus, the hypothalamus can affect any organ or part of the body.

Although studies of meditation and other spiritual experiences have not specifically observed the hypothalamus at work during these states, the results of hypothalamic activity are clearly seen in both autonomic shifts and hormonal changes observed during such states. Meditation has been shown to alter the release of hormones such as vasopressin, which helps regulate blood pressure, thyroid-stimulating hormone, growth hormone, and testosterone—all of which are controlled to varying degrees by the hypothalamus.[14] It therefore seems highly likely that something is happening in the hypothalamus during spiritual experiences and religious practices.

The Watchdog: The Amygdala

Located in the middle part of the temporal lobe, the amygdala is also one of the oldest structures in the brain and controls and mediates virtually all high-order emotional functions.[15] It is complex enough to discern and express subtle emotional nuances such as love, affection, friendliness, and distrust. It also performs an important surveillance function, made possible by the rich neural networks that connect this structure to other brain regions. The amygdala uses these connections to monitor sensory stimuli throughout the brain, searching for any input that represents the need for action—a sign of opportunity or danger, or anything else that might be worthy of the mind's attention.

When a stimulus requiring our attention is presented, the amygdala acts to analyze its significance in a very basic way, then directs the mind to pay attention by assigning emotional value to the stimulus. If you hear a suspicious noise at night, for example, it's the amygdala, via the body's arousal system, that quickens your pulse and triggers the surge of fear that makes your eyes snap open. In the case of more positive stimuli—the smell of food, for example, or the sight of an attractive sexual partner—the same process occurs, with the amygdala calling the stimulus to the mind's attention by investing it with the appropriate emotional response.

This monitoring function has been demonstrated in animal experiments, in which electrical stimulation of the amygdala results in rapid glances and searching movements of the head.[16] The animal appears anxiously expectant, its heartbeat quickens, its rate of respiration increases, and other physiological arousal effects kick in. Brain imaging studies also demonstrate increased amygdalar activity during arousal states.

The ability of the amygdala to trigger autonomic arousal activity is a key element in the generation of human emotion, but the amygdala does not exert its influence directly upon the autonomic system. Instead, it activates the hypothalamus, which in turn influences autonomic activity.[17]

The Diplomat: The Hippocampus

Located slightly behind the amygdala in the temporal lobe, the hippocampus is greatly influenced by the activity of the amygdala, and the two structures often act in a complementary way to focus the mind's attention on interesting sensory input, to generate emotions, and to link those emotions to images, memory, and learning.[18]

The hippocampus also seems to exert a regulatory effect on

another limbic structure, the thalamus. The hippocampus, both by itself and in conjunction with the thalamus, can often block sensory input to various neocortical areas.[19] The hippocampus also has the power to regulate the quiescent and arousal reactions generated by the autonomic nervous system in order to avoid extreme arousal states and maintain emotional equilibrium. Unlike the amygdala and the hypothalamus, the hippocampus does not directly generate emotion, but by its regulatory effects upon other key parts of the brain, it exerts great influence upon an individual's state of mind.

The emotional brain, its master controller, watchdog, and diplomat are all involved in creating everyday perceptions and likely are important for spiritual experiences as well. In addition, these physical structures work closely with other parts of the brain in complex ways to create the higher functions of the mind and the kinds of thoughts and views that are uniquely, individually human.

HOW THE MIND UNDERSTANDS THE WORLD: THE COGNITIVE OPERATORS

Remember the painfully slow moving robot of chapter two that is unable to repeat its trip from one side of the room to the other in order to open a door? A simple *X*, placed by scientists on the door, kept it from completing its second journey. We humans almost effortlessly look beyond the *X* to see a door.

At every moment, we brush aside distractions, sift through the clutter of mental and external stimuli, and create accurate, reliable renditions of the world outside the skull. And yet the problem faced by that primitive robot brain is the same problem faced by the human brain at its most fundamental level of interaction with the world: We need to extract meaning and relevance from the constant flood of sensory information bombarding our brains every moment of our existence.

What separates us from the robot is the nearly instantaneous communication and interaction that takes place between the different parts of the human brain. We can see these miraculous workings clearly in children. When, for example, a young child who has an orange-and-white cat sees a large black dog, she may call the dog "cat." She does not, however, call the fluffy, orange-and-white pillow on her bed "cat." If all sensory information reaching her brain were weighted equally, as it is in the brain of the robot, she would more than likely think the opposite. And yet she is able to make the distinction that the cat is more like the dog than the pillow. These rich perceptions generated by the brain are produced by the areas that make associations between the different data we receive and our analysis of them. These complex brain structures create a vivid rendition of the world around us, and allow us to interact with our surroundings effectively and confidently.

These areas are, in a figurative sense, the neurological anchors of the mind. They also support some of the "cognitive operators," the term Gene and I have created to describe the human mind's most general analytical functions.[20] The cognitive operators are not easy to describe. In simplest terms, they represent the things an effective mind is able, and inclined, to do. They are not, themselves, structures of the brain; more accurately, the cognitive operators refer to the collective functions of various brain structures. For example, if we say that the quantitative operator (the operator that deals with numbers and mathematics and makes sense out of them for our daily living) helps us solve a complex math problem, we mean that *all* the brain structures and functions involved in solving the problem are doing their job.[21]

Gene and I have identified the cognitive operators by asking a simple question: What sort of abilities would the mind require to enable us to think and feel and experience the world in a way we define as distinctively human? The cognitive operators are the answers to that question. These operators shape all our thoughts and feelings, but they are not themselves ideas. They are, instead, the organizing principles of the mind. They refer to the mind's overall function and draw on many key brain structures to do so. The cognitive operators are our general human ability to think, feel, interpret, and analyze our world. They also give us our individual intellectual stamp—they make our thoughts and emotions uniquely our own.

While there may be other cognitive operators, the eight we have named are the most relevant to our discussion of religious experience. For the sake of our discussion, we are going to look at each of these functions as an individual process. In most cases, however, the operators work in complex harmony to make us the thinking, feeling, conscious beings that we are.[22]

Seeing the Forest in the Trees: The Holistic Operator

In its basic form, the holistic operator allows us to see the world as a whole. Thanks to this mental function, we can look at an assemblage of component parts—bark, leaves, and branches, for example—and instantly, without effort, comprehend a tree. The holistic operator most likely rises from the activity of the parietal area in the brain's right hemisphere.

Seeing the Trees in the Forest: The Reductionist Operator

Rising primarily from the action of the brain's analytically inclined left hemisphere, the reductionist operator works in a way that is the antithesis of the holistic function. It enables the mind to see the whole broken down into its component parts. This is the mental function that, for example, helps us see the microclimates that are within the global environment as a whole.

The Mind's Taxonomist: The Abstractive Operator

The abstractive operator, which likely results from the operations of the parietal lobe in the brain's left hemisphere, permits the formation of general concepts from the perception of individual facts. For example, it gives us the ability to recognize a dachshund, dalmatian, and Irish wolfhound as members of a single conceptual category. Once such a category has been formed, other parts of the brain can give that category a name—in this case, "canines." Without the abstractive function, the linguistic naming of categories, and all other general concepts and ideas underlying language, would be impossible.

In an even more complex function, the abstractive operator allows the mind to find links between two separate facts. Any idea that is based on factual evidence but is not known to be factual itself is generated by the abstractive operator. Thus, the abstractive function is capable of generating scientific theories, philosophical assumptions, religious beliefs, and political ideologies.

The Mathematical Mind: The Quantitative Operator

The quantitative function permits the abstraction of quantity from the perception of various elements. The quantitative operator is obviously related to the mind's ability to perform mathematical operations, but it also performs more basic survival functions, such as the estimation of time and distance, the need to be aware of amounts (in terms of available food or the number of enemies approaching), and the need to order objects or sequences of events by some numerical system. Research suggests that the quantitative function may be genetically endowed, and that even babies younger than one year of age can understand the basic concepts of math, such as addition and subtraction.[23]

The How and the Why: The Causal Operator

The causal operator enables the mind to interpret all of reality as a sequence of specific causes and effects. It gives us the ability to anticipate and identify causes as well as to recognize that the possibility of causes even exists. The relationship between causes and effects may seem simplistic and self-evident, but nothing is ever self-evident to the mind unless it has been first processed through the neurological machinery of the brain. In fact, research shows that people who suffer damage to the parts of the brain in which lies the neurological foundation for the causal function lose the ability to easily determine the causes of even the simplest events. Such a person might be puzzled by the sound of a doorbell, for example, because the neural structures that would normally suggest to the mind that a visitor might be waiting at the doorstep have been destroyed.

The causal operator is likely the driver of almost all human curiosity. It drives us to find the causes of all things that interest or concern us, and is apparently the compulsion behind all the attempts of science, philosophy, and especially religion to explain the mysteries of the universe.

This versus That: The Binary Operator

The binary operator gives the mind one of its most powerful tools for organizing reality and allows us to move confidently and efficiently through the physical world. It enables the mind to make fundamental sense of things by reducing the most complicated relationships of space and time to simple pairs of opposites—up versus down, in versus out, left versus right, before versus after, and so on.

The existence of opposites, like the existence of causes, may seem to the practical observer a foregone conclusion. But just as the mind would not apprehend the possibility of causes without the work of the casual operator, the mind's ability to define things in terms of what they are not would be impossible without the presence of the binary function. In fact, people with serious damage to the inferior parietal lobe, which performs the neurological operations from which the binary operator rises, lose their ability to name the opposite of any object or word presented to them. Similarly, the same individual would be unable to contrast one object with another and to describe them in any kind of relative degree. He'd be stumped if you asked him to describe the difference between, say, a bowling ball and a marble, because concepts like "bigger than" and "smaller than" are simply beyond his mental reach. The binary operator helps the mind make significant distinctions in the physical and ideological world.

No Exit: The Existential Operator

The existential operator is the function of the mind that assigns a sense of existence or reality to the sensory information processed by the brain. In simple terms, this operator gives us the sense that what the brain shows us is real.

The presence of an existential operator is implied by the results of several recent studies. In one such study, babies watched a ball roll from left to right across a tabletop, and then disappear behind a screen. When the screen was lifted, the babies saw the ball resting against a wall to their right. The experiment was then repeated, and when the screen was lifted, the babies saw the ball resting against the left side of a solid box. The ball was then rolled a third time, but this time, when the screen was lifted, the babies saw the ball resting on the right side of the box. Observation showed that the babies spent a significantly longer time staring at the ball that appeared to have rolled through the box.[24]

This was interpreted as a sign that the babies somehow understood the impossibility of what they were seeing—that one solid, real thing can't pass through another—which in turn implies that even babies naturally comprehend the concreteness of reality.

The existential operator most likely resides, in part at least, in the limbic system, since emotion is such an important part of all real experiences. But our sense of things being real also requires sensory elements—we need to touch, hear, smell, taste, and, of course, see things before we can judge them as real—so it is probable that the existential operator also derives some of its function from the sensory association areas.

The Feeling of What Happens: The Emotional Value Operator

All the cognitive operators previously mentioned work to provide us with a uniquely human, sophisticated way to interpret the world. They allow us to infer cause, quantity, order, and unity in the elements that comprise the world around us, or to portray those elements as opposites, or break them down into smaller parts. Each of these functions has obvious survival value, but all of them are simply mental interpretations of our perceptions. They do not allow us to evaluate the things the brain perceives emotionally.

The emotional value operator exists to assign an emotional valence to all the elements of perception and cognition. Without the emotional value function, we would move through the world like very intelligent robots. The other mental functions would allow us to comprehend and efficiently analyze the environment, but without the motivational push that comes from feeling fear, or joy, or the longing for survival, we may not have become the successful species that we are.

Evidence for the importance of this operator comes partially from the somatic marker hypothesis outlined by Antonio Damasio in his book *The Feeling of What Happens*.[25] He proposes that emotions are critical to human reasoning and rational thinking. Without the work of the emotional value operator, we would not be driven to seek the company of others, we would not desire mates, or care for our young. By investing these important behaviors with emotional value, the brain ensures that we will pursue survival intensely and passionately.

Seeing the Brain in Action

Brain activation studies using PET, SPECT, and fMRI have given us a fairly detailed picture of the specific functions of the individual structures of the brain. We can determine which brain regions are associated with each of the five senses, which regions are activated by motor behaviors, from whole-body movements to the wiggle of a little finger. We can watch various parts of the brain turn on and off as subjects do addition and subtraction, write a letter, experience pain, or gaze upon the face of a friend.

The conclusion to be drawn from this growing fund of knowledge is that every event that happens to us or any action that we take can be associated with activity in one or more specific regions of the brain. This includes, necessarily, all religious and spiritual experiences. The evidence further compels us to believe that if God does indeed exist, the only place he can manifest his existence would be in the tangled neural pathways and physiological structures of the brain.

4

MYTH-MAKING

The Compulsion to Create Stories and Beliefs

Somewhere in the mists of human prehistory, our slope-browed Stone Age cousins, now known as Neanderthals, apparently became the earth's first living creatures to bury their dead with ceremonies. We can only imagine what dark thoughts possessed those gruff and shaggy nomads as they gently lay their clan mates to rest. What we do know is that more was going on than the simple disposal of human remains, because the graves had been carefully provisioned with tools, weapons, clothing, and other essential supplies. Perhaps these were gifts, comparable to the flowers, wreaths, carved stones, and trees we plant in memoriam today. More likely, it seems our Neanderthal progenitors were outfitting their dead with gear to help them meet whatever mysterious adventures lay ahead.

This poignant and optimistic gesture—history's first-known glimmering of metaphysical hope—tells us two important things about our Neanderthal ancestors: first, that they possessed sufficient brain power to comprehend the inescapable finality of physical death; and second, that they had already found a way to defeat or cope with it, at least conceptually.

Evidence of Neanderthal mortuary rituals has been discovered at Paleolithic gravesites scattered across Europe and Asia, and while anthropologists know very little about the specifics of Neanderthal myth, these early humans had clearly devised a system of belief that assured them that in some sense, death could be survived.[1]

The Neanderthals, it seems, had also come to believe that their world was not chaotic, but was instead governed by powerful, orderly forces that they could come to know. They believed they could appeal to these forces through proper practices and, to some extent, control them. We know this because Neanderthal shrines have been found in high mountain caves where bear skulls had been ritualistically stacked in pyramids and small crude altars still show the charred evidence of animal sacrifices carried out as long ago as two hundred thousand years.[2]

The graves and shrines of the Neanderthals are the earliest known evidence of protoreligious behavior. The fact that they occur coincidentally with the earliest evidence of human culture—pottery, complex tools, rudimentary housewares—suggests something important: As soon as hominids began to behave like human beings, they began to wonder and worry about the deepest mysteries of existence—and found resolutions for those mysteries in the stories we call myths.

"Mythology is apparently coeval with mankind," says renowned scholar of myth Joseph Campbell. "As far back, that is to say, as we have been able to follow the broken, scattered, earliest evidence of the emergence of our species, signs have been found which indicate that mythological aims and concerns were already shaping the arts and world of *Homo sapiens*."[3]

Myths are apparently as old as human culture, but it would be a mistake to write off mythical thinking as a vestige of the archaic past. Myths are alive today in the foundational stories that empower all modern religions—for instance, the story of Jesus and the saga of the Buddha's enlightenment. This is not to say that either story is necessarily untrue.

Contrary to the meaning conveyed by modern usage, the term *myth* is not a synonym for "fantasy" or "fable." It does not specifically imply falsehood or fabrication. Instead, in its classical definition, the word has an older, deeper meaning. It comes from the Greek *mythos*, which translates as "word," but one spoken with deep, unquestioned authority. *Mythos* is, in turn, anchored in the Greek term *musteion*, which, according to religion scholar Karen Armstrong, author of *A History of God*, means "to close the eyes or the mouth," rooting myth, Armstrong says, "in an experience of darkness and silence."

For Joseph Campbell, this darkness and silence lies at the heart of the human soul. The purpose of myth, he says, is to plumb these inner depths and tell us—in metaphor and symbol—of "matters fundamental to ourselves, enduring principles about which it would be good to know if our conscious minds are to be kept in touch with our own most secret, motivating depths."[4]

Myths, Campbell tells us, show us how to be human. They show us what is most important, and what, in terms of the inner life, is most deeply and profoundly true. The power of myth lies beneath its literal interpretations, in the ability of its universal symbols and themes to connect us with the most essential parts of ourselves in ways that logic and reason alone cannot. By this definition, religions *must* be based in myth if they are to have anything meaningful to say to us. In this sense, the story of Jesus is a myth even if it were literally and historically true. Likewise, even if the extraordinary events that myths chronicle never happened, and the beings they portray never walked the earth, the lasting myths of past cultures all contain psychological and spiritual truths that resonate with the psyches and spirits of readers today.

All religions, in essence, are founded upon myths. It follows, then, that any hunt for the neurological roots of the religious experience should begin with an examination of this congenital human genius for telling and believing mythic stories. We begin by asking a simple question: Why would the human mind compel us, in every culture and throughout time, to seek answers to our most troubling problems in myth? At first glance the answer might seem obvious: We rely on myths to alleviate our existential fears and comfort us in a baffling and dangerous world.

Yet the comfort that comes from these sometimes incredible-sounding stories and their reassurances may seem odd if we accept that the brain and mind evolved to enhance the individual's chances of physical survival. Why would such a practically oriented mind find reassurance in what could be a creative fabrication of its own imagination? To answer this question fully, we must understand the biological and evolutionary purpose behind the myth-making urge, and how that purpose has resulted in the neurological mechanisms that can unlock the deepest spiritual potentials of the human mind.

Death has never been a stranger in the natural world. We have no way of knowing what animals in the wild make of the grisly reminders of mortality with which they are constantly confronted. Elephants seem to have some apprehension that living things die—researchers have observed elephant families traveling considerable distances to visit, and gently caress, the skeletal remains of dead relatives. Other intelligent creatures such as whales, dogs, and monkeys also appear to mourn. But we have no reason to believe that animals dwell upon the mysteries of death. They seem more preoccupied with sensing and avoiding danger. In the ruthless animal world, threats are constant and brutally clear, and danger is a thoroughly unmysterious neighbor.

There's no hidden meaning, for example, in an antelope's flight from a cheetah, either the antelope escapes or it doesn't. If it does, and is able to rejoin its herd, it only knows, in whatever way antelopes *know* anything, that it has escaped a terrifying but specific threat. We have no definite evidence that it is capable of pondering some larger definition of death, any more than it has the capacity to wonder, as it grazes, why grass grows or what holds up the sun. For the antelope, death *is* the cheetah—nothing more and nothing less—and every antelope on the planet knows to run when the shadow of death approaches.

The biological impulse behind the antelope's urge to flee begins in the animal's limbic system, which generates arousal in response to sensory perceptions of danger—for example, the sight or scent of the cheetah. Activation of limbic structures triggers the autonomic nervous system, which in turn augments this arousal response by releasing adrenaline, raising heart rate and respiration, increasing muscle tone, and generally preparing the antelope for action.

The limbic and autonomic systems of animals like the antelope are similar to the corresponding systems in human beings, as are the way our systems activate the arousal response. The crucial difference is that in animals, the fear response is largely stimulus-bound. In other words, the full-blown fight-or-flight response is triggered only as long as the perceived threat is present. Lacking the complex brain structures required for abstract thought, animals like the antelope can't anticipate, in the abstract, the presence of a cheetah. If a grazing antelope is startled by a rustling in the brush, for instance, it may gaze anxiously in that direction as its autonomic system begins to prepare it for flight. But unless the worrisome stimulus continues, or until the actual presence of a predator can be confirmed, the levels of arousal activity do not reach the levels required to compel the animal to run.

This binding of a behavioral response to the presence or absence of a specific stimulus gives the antelope a relatively limited set of options: If the stimulus does not persist, the neurological activity of the limbic and autonomic systems subsides and the animal feels safe enough to continue grazing. If the stimulus persists, or if the predator is actually detected, then the arousal reaction surges, and the animal is neurologically obliged to run, or even fight if necessary.

The fear response in humans also involves the activation of the autonomic system via the stimulation of the limbic structures, but the complexity of the human brain adds a new and significant wrinkle. Thanks primarily to the presence of the cerebral cortex, which enables us to perform high-level cognitive operations, the human mind is able to do what animal brains cannot—think of danger in the abstract, anticipate the possibility of danger even when no threat is immediately at hand. Because the cortical structures are so intimately linked to the more primitive functions of the limbic and autonomic systems, humans are able to trigger a biological fear response simply by thinking of danger.[5] A bushman walking through lion country, for example, will feel some degree of arousal even if no lions are in sight, while the grazing animals around him might feel perfectly and thoughtlessly at ease.

Because of this knowledge of the potential threats all around them, early humans likely saw the world as a complex, endlessly dangerous place. As early humans learned more and more about the nature of the physical world, they could not help but dwell upon the hazards it presented. They had to contend with animal predators as well as human enemies. There were floods, droughts, sickness, and famine. These challenges to their existence were enough to keep them in a continual state of fearful arousal.

Thankfully, the same big brain that generated these fears also provided a way to resolve them through invention. Humans developed tools, weapons, and simple technologies. They banded into groups, allowing them to hunt cooperatively, share resources, and more efficiently defend themselves from hostile outsiders.[6] They also invented ideas to protect themselves—laws, cultures, religions, and science, which enabled them to adapt more and more to their world. All of the lofty reaches to which human achievement has carried us—from the first flint spearhead to the latest innovation in heart transplant surgery—can be traced to the mind's need to reduce the intolerable anxiety that is the brain's way of warning us that we are not safe.

The high-level thought processes that allowed human beings to perceive complex threats and resolve them in creative, sophisticated ways are what we have referred to as the cognitive operators. These general analytical functions of the mind allow us to think, feel, and experience the world in an essentially human way. These mental attributes have allowed our species to adapt creatively and successfully to even the most hostile habitats on earth.

The functions associated with these operators evolved as standard equipment in every human brain because of the adaptive advantages they offer. So effective were these cognitive powers, in fact, that evolution appears to have provided the human brain with a biological compulsion to use them. Gene and I have referred to this involuntary mental drive as the *cognitive imperative*;[7] it is the almost irresistible, biologically driven need to make sense of things through the cognitive analysis of reality.[8]

Researchers have provided support for the existence of the cognitive imperative by showing that the mind, when confronted with an overwhelming flow of sensory information, reacts with increasing anxiety. The researchers concluded that this anxiety was caused by the frustration of the mind's insatiable need to sort confusion into order and the difficulty in doing so when overwhelmed by information.

There is a simpler and more compelling way to demonstrate the existence of the cognitive imperative: glance around and try not to perceive a cohesive portrait of your world. Simpler still: try not to think. As any novice meditator knows all too well, the mind just isn't made that way.

The cognitive imperative drives the higher functions of the mind to analyze the perceptions processed by the brain and transform them into a world full of meaning and purpose. By doing so, it has given human beings an unsurpassed genius for adaptation and survival. But these cognitive abilities have a down side as well. In its tireless quest to identify and resolve any threat that can potentially harm us, the mind has discovered the one alarming apprehension that can't be resolved by any natural means—the sobering understanding that everyone dies.

This grim discovery must have entered the world soon after self-awareness began to glow in some prehistoric human mind. The moment it did, the cognitive imperative would have driven the mind to find a resolution. The problem would have engaged the cerebral cortex in the manner of any abstract thought and, soon, the limbic and autonomic systems would generate an arousal response. The intensity of the anxiety produced by such a response might not be as sharp as the response generated by a more acute concern—an earthquake, for example, or a tiger about to pounce—but as long as it persists, the cognitive imperative will continue to bring the analytical power of the mind to bear upon it.

But death was not the only existential worry that early humans had to face. By comprehending their own mortality, they had stumbled onto a new dimension of metaphysical worries, and their questioning minds must have presented them with difficult and unanswerable questions at every turn: Why were we born only eventually to die? What happens to us when we die? What is our place in the universe? Why is there suffering? What sustains and animates the universe? How was the universe made? How long will the universe last?

And, most pressingly: How can we live in this bafflingly uncertain world and not be afraid?

These are confounding questions, but the cognitive imperative cannot let them lie, so it tirelessly pushes the mind to find resolution. For thousands of years in cultures around the globe, that resolution has been found in the form of myth. Myths, in fact, always begin with the apprehension of some metaphysical problem that is resolved in the mythic story through using metaphorical images and themes—Eve eats the apple; Pandora opens the box. By learning these stories, and passing them on, our questions about suffering, good and evil, and numerous other metaphysical problems suddenly become answerable, knowable.

Essentially, all myths can be reduced to a simple framework.[9] First, they focus upon a crucial existential concern—the creation of the world, for example, or how evil came to be. Next, they frame that concern as a pair of apparently irreconcilable opposites—heroes and monsters, gods and humans, life and death, heaven and hell. Finally, and most important, myths reconcile those opposites, often through the actions of gods or other spiritual powers, in a way that relieves our existential concerns.

For example, consider the myth of Jesus. As the myth begins, the world is lost in sin, and heaven is unattainable. The mythic opposites in the story are clear: a distant God and a suffering humanity. Jesus resolves these opposites in several ways. First, as the son of God in human form, He resolves them in His person; second, through His death and resurrection, He unites God and man in the promise of eternal life. Buddha provides a similar "salvation" by showing that the pursuit of enlightenment and the practice of detachment and compassion can help us understand the endless cycle of human suffering and reunite us with the sublime oneness that is our true being.

The same themes echo in the mythologies of the ancient world, in the stories of gods and heroes whose deaths and resurrections symbolically heal the rift between heaven and earth, including the Egyptian Osiris, the Greek Dionysus, the Syrian Adonis, and the Mesopotamian Tammuz.

Virtually all myths can be reduced to the same consistent pattern: identify a crucial existential concern, frame it as a pair of incompatible opposites, then find a resolution that alleviates anxiety and allows us to live more happily in the world. Why should this be so? We believe myths are structured in this way because the mind makes sense of mythical problems using the same cognitive functions it relies upon to make fundamental sense of the physical world.

The creation of complex mythic stories requires the creative, combined interaction of all the cognitive operators, but two of the operators appear to play especially significant roles. The first is the causal operator, which should be no surprise, since myths are essentially about the root causes of things. The causal operator, you'll recall, is the mind's ability to think in terms of abstract causes—to link that chuffing roar in the distance with the likely presence of a lion, for example, or to trace the pain in your belly to the unfamiliar berries you sampled last night. In the moment-to-moment flow of conscious thinking, we take such causal associations for granted, but the mind would not possess the potential to understand the concept of cause without the analytical powers of the causal operator. Nor would it be able to create the many stories about creation.

The second cognitive operator crucial to the myth-making mind is the *binary operator*, which refers to the brain's ability to frame the world in terms of basic polar opposites. The human brain's ability to reduce the most complicated relationships of space and time to simple pairs of opposites—above and below, in and out, left and right, before and after, and so on—gives the mind a powerful method of analyzing external reality.

Again, we tend to take this crucial mental process for granted; after all, what could be more obvious than the notion that the opposite of "up" is "down"? But the relationship between "up" and "down" is not as absolute as it seems. In fact, it's really quite relative and arbitrary and only feels obvious to us because our minds have evolved to see things that way.

In other words, the binary operator does not simply observe and identify opposites, but in a very real sense it creates them, and it does so for an evolutionary purpose. In order to negotiate the environment confidently, we need a way to divide space and time into more comprehensible units. Relationships such as above and below, inside and outside, before and after, and so on, give us a basic way of orienting ourselves to the outside world, of *feeling* our way through the environment.

These relationships are conceptual, of course, and far from absolute; "up" for example, would have very little meaning to an astronaut far from earth. But the cognitive processing of the binary operator turns these relationships into something tangible and absolute, and so makes better sense of the physical world. So, when the cognitive imperative, driven by some existential fear, directs the binary function to make sense of the metaphysical landscape, it obliges by interrupting that existential problem and rearranging it into the pairs of irreconcilable opposites that become the key elements of myth: heaven and hell; good and evil; celebration and tragedy; birth, death, and rebirth; isolation and unity.

THE GENESIS OF MYTH

As we've already mentioned, the earliest physical evidence of mythological activity dates back to the Neanderthals, whose brains, while somewhat less sophisticated than the brains of *Homo sapiens*, likely contained the neurological structures that could support causal and binary function. It's possible, though, that myth-making originated with even earlier hominid species, and that the traces of that activity have simply been eradicated by time. Our understanding of prehuman brains, incomplete though it is, suggests that this might be the case, that the need to perceive and understand the metaphysical aspects of existence surfaced very early in the emergence of the human family line.

Whoever the first myth-makers actually were, they were likely set apart neurologically from the rest of creation by the presence and function of a well-developed parietal lobe. In humans, the parietal region contains the neurological structures that support the causal and binary operators, as well as the center of language, which is also necessary for the formation of myth. In a sense, the parietal lobe is a crucial part of the brain's myth-making center. No brain lacking some kind of parietal area could think in terms of opposites, so it could not create the basic components of mythic structure. Nor could it understand the concept of cause, most likely eliminating the need to make myth at all. Conversely, any brain equipped with these abilities would be driven to use them to analyze all of its experiences. When focused upon the unanswerable mysteries of existence they would inevitably resolve these mysteries through myth as has been demonstrated in every human culture and even in most human ancestors with this brain apparatus.

A rudimentary version of the parietal lobe is present in our close evolutionary relative, the chimpanzee. While chimps are smart enough to master simple mathematical concepts and develop nonverbal language skills, their brains appear to lack the neural complexity needed to formulate any significant kind of abstract thought, which is the type of thought that leads to the formation of cultures, art, mathematics, technology, and myths.[10]

A more developed parietal lobe appeared in *Australopithecus*, an early, apelike biped human ancestor who foraged the wilderness several million years ago. Plaster casts of australopithecine skulls show a parietal area that, though small, had developed enough to perform rudimentary conceptualizations, which would enable thinking in terms of opposites, and understanding of the concept of cause.[11] These qualities would make *Australopithecus* the first of our ancestors with the minimal mental machinery for generating myth. The chances are, however, that they never did so. Evidence shows that in the australopithecine brain, the parietal area was not supported by the neural structures needed to enable language and verbal speech that also appears necessary for the full development of myth.[12] Lacking the ability to speak, or even think, in verbal language, *Australopithecus* may have been able to feel existential dread, and even intuitively grope toward a resolution. Perhaps such a mind could even construct a personal mythology based entirely on abstract, nonverbal symbols, but the possibility is remote.

Other versions of the parietal lobe appeared in distant primate species that preceded the evolution of the human line, but none of them measures up neurologically to the needs of the myth-making mind. In fact, the first mind truly capable of creating myths probably does not emerge until the advent of the genus *Homo*, our own nuclear family, in evolutionary terms. That mind belonged to a tool-making hunter-gatherer known as *Homo erectus*, who walked upright several hundred thousand years ago. Inside the relatively roomy vault of *Homo erectus*'s skull was a complex brain containing all the key neural structures for creating language and speech.[13] Also present was a well-developed parietal lobe, with all the relevant connections and convolutions, which almost certainly granted him the myth-making powers of causal and antinomic thinking.

We don't know for certain whether *Homo erectus* ever used his myth-making potential—he left no physical evidence of ritualized behavior. Such would not be seen until the time of the Neanderthals, almost a hundred thousand years ago. But the presence of causal and binary functions, and the clear ability to form language and speech, make a strong case for the argument that *Homo erectus*, one of the founders of our family line and the first living creature we would reasonably call human, was also the first sentient being likely to perceive a spiritual reality—a realm of beings and forces beyond the material world—and to define that reality through the stories of myth.[14] To fully understand how, and why, the mind's cognitive functions would biologically compel him to do so, we must first understand how they resolve dilemmas of a more immediate and tangible kind.

Imagine, for example, a prehistoric hunter making his way home through unfamiliar woods. His mind wanders as he travels, and he is only absently aware of the ambient noises of the forest, but when a twig snaps in the underbrush, his mind is instantly, involuntarily focused. This intense mental alertness results from the sudden activation of the amygdala, the ancient limbic watchdog that monitors all incoming sensory information for signs of danger and opportunity. When the amygdala detects the unaccounted-for auditory impulses caused by the sudden noise, it rivets the hunter's mind upon it. Already, the autonomic system has triggered an arousal response, bracing the body for action. In the same split second the hunter first hears the suspicious sound, the cognitive imperative is driving the causal operator to discover what it might mean.

Discovering a cause is the first priority of the causal operator, but as the hunter scans the underbrush, no cause is apparent. Uncertainty at such an urgent moment is intolerable to the causal operator, so it does what it is designed to do in the absence of a specific cause: it proposes one. This proposal arises out of activity in the hippocampus, the limbic structure where past experiences are stored as memories. Rapidly, the mind scours these memory banks, sorting and cross-referencing information as it searches for any pertinent content—images, sounds, or larger chunks of experience—that might shed light on the problem at hand.

The computational task is staggering, but in an instant, all the brain's memory files have been consulted, all irrelevant data has been ignored, and the causal operator presents its best hunch by conjuring in the hunter's mind the idea of a leopard lurking in the trees. For the hunter, who had seen the tracks of a big cat earlier that day, and who had once been chased by a leopard in woods very similar to these, there was no need for further deliberation; the unseen predator was utterly real, the danger was immediate, and his only option was to run.

Later, in a calmer moment, the hunter might reconsider his

reaction. After all, he had no way of knowing for sure that the noise was made by a leopard—it's just as likely that the twig was stepped on by a plump wild pig, or by a harmless foraging deer. But the hunter didn't need to *know* the leopard was real, it was enough that he *believed* it. The causal operator is designed to promote survival, not necessarily to find the truth, and if there really had been a leopard in the bushes, it was the hunter's ability to *think* about the potential for danger, as opposed to an antelope that can only *react*, that would save his life.

But why would he believe so surely in something he didn't know absolutely to be true? We think there was more to the hunter's reaction than simple common sense. We believe he accepted the realness of the unseen leopard because neurological forces allowed him no other choice.

The hunter's process of believing that began as amygdalar function focused the mind's cognitive operations upon the mystery of the noise in the bushes. The causal operator, or, more specifically, the brain structures that support the causal operator, responded by proposing the presence of a leopard. Simultaneously, the binary operator interprets the problem as a conflict of opposites. In a specific sense, this conflict is between the concepts of leopard and not-leopard, but on a deeper and more general level it is a fundamental conflict of life versus death.

In either case, these urgent conflicts must be effectively resolved. The analytical, verbally inclined left brain immediately attacked the problem by making certain logical connections: The hunter realized he was in leopard country, and remembered having seen leopard tracks a few miles back. It was logical to believe that a leopard might have been lurking in the trees.

At the same time, the hunter knew that the tracks he'd seen were at least several days old. He also knew that leopards do not customarily hunt at that time of the day. This led to the logical possibility that something other than a leopard—a deer or a pig, for example—was the source of the startling sound. The hunter now faced a logical dilemma: Fleeing might have cost him an easy kill, but hesitation might have cost him his life.

While the verbal, analytical left brain struggled with this problem, the intuitive, holistic right brain was taking a different approach. Thinking with images and emotions, rather than language and logic, the right brain analyzed how the situation felt. The right brain visualized the easy kill, and the reaction was highly positive. But those positive feelings were heavily outweighed by the vivid visualization of being eaten by a leopard. As the right brain dwelt upon this grisly possibility, a memory sprang to mind of a time when the hunter was chased by a man-eating cat in woods very similar to these. The right brain remembered the terror and instantly made its decision: There's a leopard in the trees.

This emotional apprehension immediately colored the decision-making process of the left brain. The logical idea of the leopard was now charged with emotional substance, and as the intuitive faculties of the right brain came into unified agreement with the logical powers of the left, the ideas in his head gained depth and authority. He not only thought there was a leopard in the bushes, he felt it in his bones.

Now the opposites of leopard versus not-leopard, and of life versus death, had been powerfully and neurologically resolved. A cause had been determined. Ideas had become emotionally charged convictions; a logical possibility had become a visceral belief.[15]

In a sense, the hunter had created a simplistic myth—the myth of the leopard in the bushes. It began, as all myths do, with urgent, unanswerable questions. In this case: What is that noise, and what does it mean? The importance of finding answers compelled the mind, through the drive of the cognitive imperative, to bring the brain's analytical powers to bear. The causal operator found a plausible explanation for the noise. The binary operator framed the problem in terms of opposites. And finally, the holistic agreement of the left and right sides of the brain led to a whole brain unification that turned logical ideas into emotionally felt beliefs. These beliefs resolved all uncertainties and gave the hunter a coherent scenario in which he could react effectively.

Like all effective myths, the story of the leopard in the trees may or may not have been literally true. Yet this simple scenario explains the unexplainable in a way that allowed the hunter to take effective and possibly life-saving action. His belief enhanced his chances of survival, and that is precisely the goal of the cognitive drive.

This process is automatic: uncertainty causes anxiety, and anxiety must be resolved. Sometimes resolutions are obvious and causes are easy to spot. When they are not, the cognitive imperative compels us to find plausible resolutions in the form of a story, like the story of the leopard in the trees. These stories are especially important when the mind confronts our existential fears. We suffer. We die. We feel small and vulnerable in a dangerous and confusing world. There is no simple way to resolve these enormous uncertainties. In such situations, the explanatory stories that the mind creates take the shape of religious myth.

It would be impossible to trace the endless web of cultural and psychological factors that lead to the genesis of a single specific myth, and it would be foolhardy to suggest that anyone could explain for certain how any specific religious myth developed. But if we frame our discussion carefully, we can certainly examine the biological origins of the urge to make myth. We can even speculate upon the neurological origins of a single mythic concept. Consider, for example, the following scenario:

In a close-knit prehistoric clan, one of the tribe has died. His body lies on a bearskin. Others approach and touch him gently. They sense immediately that the man who used to he exists no more. What was once a warm and vital person has suddenly become a cold and lifeless thing.

The clan's chieftain, an introspective man, slumps beside the campfire and broods upon the lifeless form that was once his comrade. What is it that is missing? he wonders. How was it lost and where has it gone? As he watches the crackling fire his stomach tightens with sadness and anxiety. His mind's need for cause will not rest until it finds resolution, but the longer he dwells upon the unnerving puzzle of life and death, the deeper he sinks into existential dread.

In neurobiological terms, the grieving chief is in the grip of the same arousal responses that seized the startled hunter. It began in the chieftain's brain when the amygdala noticed frustration in the thought processes of the logical left hemisphere—the effect of the chief's intense, prolonged brooding. Interpreting this frustration as a sign of distress, the amygdala triggered a limbic fear response, and sent off neural signals that activated the arousal system. Now, as the chief continues to ruminate upon his grief and fear, that arousal response intensifies. His pulse quickens, his breath grows shallow and rapid, and his forehead becomes beaded in sweat.

The chief stares vacantly at the fire, turning his troubles around and around in his head. Soon the fire has burned down to embers, and as the last flames sputter and die, an intuition strikes him: The fire was once bright and alive, but now it's gone, and soon there will be nothing but lifeless gray ashes. As the last wisps of smoke rise to the heavens he turns to the body of his fallen friend. It occurs to him that his comrade's life and spirit have vanished as completely as the flames. Before he can consciously phrase the thought, he is struck by the image of the very essence of his friend escaping to the heavens, like smoke, the rising spirit of the fire.

This conviction begins as nothing more than an idea, just one more possibility offered up by the intellectual pondering of the brain's left side. At the same time, the brain's right side is proposing holistic, intuitive, nonverbal solutions to the problem. As the intellectual idea of the ascension of the spirit enters the chief's consciousness, it becomes "matched" with one of these emotional right-brain solutions. Suddenly, the agreement of both sides of the brain causes a neurological resonance that sends positive neural discharges racing through the limbic system, to stimulate pleasure centers in the hypothalamus. Because the hypothalamus regulates the autonomic nervous system, these strong pleasure impulses trigger a response from the quiescent system, which the chief experiences as a powerful surge of calmness and peace.

All this happens in the wink of an eye, too fast for the arousal response that triggered the chief's anxiety to subside. For a remarkable moment, both the quiescent and arousal systems are simultaneously active, immersing the chief in a blend of fear and rapture, a state of intensely pleasurable agitation that some neurologists call the Eureka Response, which the chief experiences as a rush of ecstasy and awe.

In this transforming flash of insight, the chief is suddenly freed from his grief and despair; in a deeper sense, he feels that he has been freed from the bonds of death.

The insight strikes him with the force of revelation. The experience feels vividly, palpably real. In that moment, the opposites of life and death are no longer locked in conflict; they have been mythically resolved. Now he sees clearly the absolute truth of things—that the spirits of the dead live on.

He feels that he has discovered a primal truth. It is more than an idea, it is a belief he has experienced in the deepest reaches of his mind.

Like the story of the leopard in the bushes, the chief's intuition about the ultimate destiny of the soul may or may not be true. What matters is the notion that it is based on something deeper than imagination or wishful thinking. We believe that all lasting myths gain their power through neurologically endorsed flashes of insight, like the one that struck the chief. These insights might take many forms, and can be triggered by many different ideas. For example, the chieftain might have seen fog rising up a mountain slope in the moonlight, and concluded that like the eerie mist, the spirits of the dead go off to dwell in the holy hills. Any idea might trigger a myth if it can unify logic and intuition, and lead to a state of left-brain/right-brain agreement. In this state of whole-brain harmony, neurological uncertainties are powerfully alleviated as existential opposites are reconciled and the problem of cause is resolved. To the anxious mind, this resonant whole-brain agreement feels like a glimpse of ultimate truth. The mind seems to live this truth, not merely comprehend it, and it is this quality of visceral experience that turns ideas into myths.

That personal myth becomes a communal myth when it is shared with others who find meaning and power in the resolutions it provides. There is no guarantee that this will happen. The chief's clan mates, for example, will not accept his insights unless, as they listen to him share his story, they experience within themselves the same neurological ring of truth that seized the chief during his moment of penetrating vision. Their reactions, in hearing the story, will not necessarily be as intense as the chief's, but if they feel even a small degree of neurologically experienced emotion, the chief's passionate testimony will gain credence. They will believe him not because they think he is right, but because they feel it. The chief may become considered a seer, and a system of mythology may rise from his beliefs.

In this scenario, the creation of mythologies is a two-step process. First, a neurologically inspired flash of insight empowers a story with the authority of myth; second, the sharing of the story triggers a similar, and usually milder, flash of insight in whoever hears the tale.

This leaves us with a pair of obvious but intriguing questions: Why, of all ideas, would the notion of the soul rising to the heavens strike such a strong holistic resonance in the grieving chieftain's mind; and why would the same idea resonate at all in the minds of the others? Or, in a broader sense: Why are the myths of all world cultures so strikingly, consistently similar? The widely acclaimed work of Joseph Campbell and other scholars of myth makes it very clear that in every human culture, across the span of time, the same mythological motifs are constantly repeated: virgin births, world-cleansing floods, lands of the dead, expulsions from paradise, men swallowed down into the bellies of whales and serpents, dead and resurrected heroes, the primeval theft of fire from the gods. . . .

In theme, detail, and intention, these stories are often remarkably alike. For example, the gospels tell us that Jesus spent forty days exiled in the desert, fasting and praying and enduring the temptations of Satan who hoped to break Jesus' faith and deter him from his redemptive destiny. Jesus survived his ordeal and returned to the world a transformed man, ready to begin the mission that would lead to his death on the cross and subsequent resurrection, which would open the doors to heaven and restore the gift of everlasting life.

In Buddhist scripture, the young prince Siddhartha sat forty days exiled in the wilderness, fasting, meditating, and enduring the temptations of the demon Mara, who hoped to distract him from his meditation and deter him from his world-changing fate. The prince survived his ordeal in the wilderness to emerge from his epic meditation as a transformed being who, by "dying" to the world of the flesh and being reborn as a pure, enlightened spirit, taught the world how death and suffering can be truly understood through liberation from attachment to the material world.

Cultural dispersion can account for some of these similarities—myths were often borrowed from culture to culture and then shaped to fit the local needs. But even if the themes and symbols of myths had been borrowed and carried to every corner of the globe, what gives them such universal power that the same essential stories would be as appealing to people in environments as different as those inhabited by Inuits, Hebrews, Incas, and Celts?

Carl Jung believed that myths were the symbolic expression of archetypal ideas—inherited forms of thought that exist, in universal form, in the depths of every human mind.[16] Joseph Campbell agreed with Jung that myths were expressions of these basic structural components of the mind. He believed, for example, that it was the influence of these deep structures, these mental archetypes, that led ancient architects, in far-flung corners of the world, to conceive the similar shapes and proportions of Summerian ziggurats, Mayan pyramids, and step-shaped Buddhist temples.

According to Campbell, the interpretation of archetypes is eventually shaped by many factors—geography, cultural needs, even the characteristics of local plants and animals. But the essential, archetypal forms and ideas remain remarkably the same. They cannot help but be similar, because they are shaped by such unchanging aspects of the mind.

"[Myths]" Campbell says, "are telling us in picture language of powers of the psyche to be recognized and integrated into our lives, powers that have been common to the human spirit forever, and which represent that wisdom of the species by which man has weathered the millenniums."[17]

Whether or not the archetypes described by Jung do indeed exist, we agree that myths are created by basic, universal aspects of the brain, in particular, the fundamental neurological processes through which the brain makes sense of the world. Although culture and psychology may influence them significantly, it's the neurological grounding of mythic stories that gives them their staying power, as well as the authority through which they resolve our existential fears.

5

RITUAL

The Physical Manifestation of Meaning

At midnight, in the shadowy choir loft of a candlelit gothic cathedral of the Calvary Episcopal Church in Pittsburgh, a fifty-four-year-old businessman named Bill sits in a crowded pew enjoying a concert by the innovative jazz ensemble known as the Paul Winter Consort. It's a hallmark of Winter's group to set their stage in unusual, and atmospheric venues—canyons, beaches, old stone barns—to reflect the moody, reverent spirit of their music, which often blends their own live performances with the recorded songs of nature. Tonight's concert, for example, has included a lyrical duet with a school of singing humpback whales and a haunting serenade built around the keening of eagles. Now, as the evening draws to a close, Winter and his group are providing the instrumental accompaniment to the tape-recorded singing of a pack of free-roaming wolves.

The rhythmic, otherworldly wolf serenade echoes eerily in the monumental quiet of the cathedral's soaring spaces. The wolves raise their voices in raw howls of sheer animal power, then let them soften into haunting and melancholy sighs. This grand old church has never heard such a choir, and the unhurried rise and fall of Winter's moody soprano sax—sometimes harmonizing with the animal voices, sometimes joining them in a gentle exchange of rhythmic call and response—adds a powerfully hypnotic dimension to the music. In a more conventional venue, the effect of this remarkable performance would weave a powerful spell, but in this old cathedral, lit only by the soft glow of candles, with shadows gathered in the eaves and dancing on the limestone walls, it's enough to lift listeners out of their everyday lives, and into another world.

And as the wolf serenade reaches its emotional crescendo, that's exactly what's happening to Bill. Quietly, unconsciously, he has allowed himself to be absorbed into the song of the wolves, lulled by its haunting rhythms and the beauty of those wild voices. He feels deeply, serenely at peace. Then, suddenly, he is seized by a surge of excitement. It rushes up from the gut in a burst of joy and energy, and before he can think twice about it, Bill is on his feet, with his head thrown back, and he is howling from the bottom of his soul.

Remarkably, at the same moment, other people have begun to howl. At first it's half a dozen, scattered throughout the church. But in moments others follow their lead and soon the entire cathedral is alive with joyous noise, as hundreds of people joyfully join in the primal song of the wolves.

"I don't know why that happened," says Bill, looking back upon the moment. "There was no signal that I can recall. I don't remember what my thoughts were, it just seemed to rise up through my body and I just let it out."

Normally a reserved and dignified person, Bill says he wasn't worried, as he howled, that others might think him strange. "I felt very comfortable with that group of people," he says, with a shrug. "Somehow, I knew I was among people who would understand." Bill doesn't know why he felt such a sense of connection with that group of strangers, or what, exactly made him howl. "It was very primal, very liberating," says Bill. "When everyone was howling, the whole church, that was a very a spiritual feeling. Not *religious*, but definitely spiritual. It's hard to put into words," he says, "there's really no way to explain it."

In fact, we believe that there is an explanation to this story; we believe that Bill and his fellow audience members were swept away by a distinct neurological chain of events that, for a few remarkable moments, relieved them of the notion that they were each aloof and independent selves, and plunged them into a sense of primal, liberating unity, not only with the wolves but with one other.

That neurological mechanism was coaxed into operation by the effects of the wolf concerto together with the rhythmic music. Research reveals that repetitive rhythmic stimulation, like the mesmerizing call of the wolves, can drive the limbic and autonomic systems,[1] which may eventually alter some very fundamental aspects of the way the brain thinks, feels, and interprets reality. These rhythms can dramatically affect the brain's neurological ability to define the limits of the self. The stimulation of autonomic and limbic responses, triggered by the rhythms of the wolf music, is the force that called Bill and his fellow audience members to rise out of themselves and into a larger and more exhilarating state of being.[2]

The modest sense of self-transcendence that Bill experienced at the concert was not specifically religious. In its ability to lift him out of the singleness of self and into a state of indefinable unity, however, it had a spiritual dimension. By creating a palpable sense of union among the audience members and by encouraging, in varying levels of intensity, emotions of tranquillity, ecstasy, and maybe even awe, the wolf concert was more than an evening's entertainment. The concert was a ritual, fulfilling both the traditional and neurological definitions of the term.

RITUAL AND UNITY

It might be difficult to think of a jazz concert, even this very unusual concert, and even a concert held in a church, as anything like a formal ritual. Yet the hypnotic rhythms of the music, the repetitive rise and fall of the voices of the wolves, even the play of candlelight and shadow on the towering walls made the wolf concerto every bit as ritualistic as any ceremonial High Mass ever held in that old stone cathedral. More important, by transforming an audience of strangers into a unified congregation, the wolf music effectively fulfilled the goal of virtually every ritual ever performed—to lift participants out of their isolated individual sensibilities and immerse them in something larger than themselves.

The transcendence of the self, and the blending of the self into some larger reality, is a major goal of ritualized behavior.[3] In a religious context the transcendent aim of ritual is to unite worshipers spiritually with a higher sense of reality. The goal of the contemplative rituals practiced by some Catholic mystics, for example, is to attain the state of the *Unio Mystica*, the mysterious union that the mystic experiences as a sense of union with the actual presence of God. In Buddhism, the aim of meditative rituals is to encounter the ultimate oneness of everything by defeating the limiting sense of self generated by the ego. Few people ever attain these lofty states, of course. For most worshipers, ritual provides a much milder degree of transcendence—the uplifting sense of togetherness shared by the members of a congregation as they finish a hymn, for example, or the sense of an intimate closeness to Jesus, felt by individuals during the ritualistic rhythms of the Catholic mass.

Religious rituals have occurred, in virtually every human culture, with an almost infinite variety of form. But in every known case, a singular principle appears to hold true: When religious ritual is effective, and it is not always effective, it inclines the brain to adjust its cognitive and emotional perceptions of the self in a way that religiously minded persons interpret as a closing of the distance between the self and God.

Not all rituals are religious, of course. Our lives are shaped by countless ritualized activities and ceremonies that are purely social or civic in nature, including political conventions, inaugurations and coronations, judicial proceedings, holiday traditions, courtship, and even sporting events. These structured ceremonies may have no religious connotations, but like all rituals they contain elements of rhythm and repetition, and they aim to define the individual as part of some larger group or cause. These secular rites show us ritual in a more practical incarnation, as a mechanism for promoting social cohesion by encouraging individuals to set aside personal interests and commit themselves to the broader interests of the common good. The social benefits of ritual, in fact, may be a major reason that ritualized behavior evolved in the first place.

THE EVOLUTIONARY ROOTS OF RITUAL

Anthropologists have long understood that the rituals of early human societies performed an important survival function by fostering, among a given clan or tribe, a sense of specialness and common destiny.[4] Through the powers of ritual, clan members were constantly reminded that they were the people favored by the particular deity they worshiped, and that they were, in some sense, chosen. This sense of special destiny set them dramatically apart from other clans, strengthened the bonds between them, and helped maintain a stable group identity as the clan's membership changed over time. As a result, the clan functioned more cooperatively and successfully, and its members enjoyed important survival advantages—protection from enemies, the sharing of resources, the creation of rules and laws—that were not available to them as isolated individuals.

Few observers would deny the evolutionary advantages provided by the phenomenon of human ritual, which would imply that ritual behavior might have biological roots.[5] But until as recently as the 1970s, researchers virtually ignored this possibility. For most of the modern era, ritual was believed to be a purely cultural phenomenon, a product of social conditioning, and not specifically biological. Consequently, little effort was made to investigate the physiological aspects of human rituals.

In the last three decades, thanks in part to the work of Eugene d'Aquili and his colleagues Charles Laughlin and John McManus, the biological side has become an important component in the study of ritual.[6] As a result, researchers have gathered considerable evidence of a link between ritual and evolution, and observation has revealed that human and animal rituals are similar in significant and surprising ways.

In their basic form, for example, animal rituals are comprised of structured, patterned responses—dancing, vocalizing, head-bobbing, and so on—that are rhythmic and repetitive in nature. These behaviors are often very dramatic and very odd, and outside the context of ritual may have no practical function whatsoever. Their purpose, it seems, is to define ritual activity as something very different from normal behavior, which sends the message that when engaged in ritual activity, an animal is up to something special.

The rites and ceremonies of human beings clearly echo the ritual behavior of animals.[7] Rhythm and repetition, for example, are elements in almost all human rituals, from the stately cadence of Gregorian chants to the livelier rhythms of a Polynesian fertility dance. Human rituals also feature distinctive behaviors—bows, slow processions, portentous gestures of the hands and arms—that serve no practical purpose in everyday life. According to biologists, human and animal rituals also serve similar and very important functions by reducing acts of aggressiveness between the members of a group, and by creating among that group a strong social bond.

These consistent similarities suggest that the rituals of animals and human beings have common evolutionary origins. They also make the case that at the simplest level, ritual behaviors evolved as a fundamental form of communication, intended to help individual animals recognize and understand the behavior of other individuals who might be significant in their lives, and whose behavior might enhance or reduce their own chances of survival.

At even relatively primitive levels, ritual behaviors seem to help animals convey their own intentions, and to interpret the intentions of other animals around them, which can help avoid misunderstandings that could result in conflict.[8] Greeting and grooming rituals, for example, allow individual animals to announce that their intentions are friendly, while rituals of submission allow subordinate animals to "show their respect" for more dominant members of the group, and keep the social hierarchy intact. These behaviors reduce tensions, enhance an animal's personal safety, and help maintain a stable social balance in the group. These behaviors also occur frequently in human life—handshakes, hugs, and the giving of gifts are often a part of human greeting rituals, for example, while social convention and good sense teach us to act deferentially when dealing with teachers, bosses, and other authority figures.

In their most primitive form, the communicative powers of ritual give even the simplest creatures the ability to send and receive messages that allow important interactions to occur. Consider, for example, the mating ritual of the butterfly known as the silver-washed fritillary.[9] When the male of this species spots a possible mate, he closes in with amorous intent, which causes the female to take to the air. As she lifts off, the male flies several looping circles around her, his wings almost brushing her body. The two then perform a spectacular joint flight, with the male darting acrobatically above and below her, as she glides, approvingly, on a straight, steady path. When the flight is completed, the butterflies land, posture, and exchange scents. In all, the male must perform seven distinct acts, and the female must respond appropriately to each one, before their mating can begin.

Virtually all species of animals perform some variation of a mating ritual. In higher animals, the courtship process may have some definable purpose—displays of agility and physical prowess, for example, might allow the female to judge the desirability of a mate. But why should creatures as simple as the butterfly indulge in such odd behavior? What could a ritual of courtship possibly mean to them?

The answer, it seems, is that for butterflies, ritual is useful only when stripped down to its bare neurobiological essence. The butterfly mating ritual is a dance of sheer neurological information, designed to identify each animal clearly as a suitable mate for the other, and to announce unambiguously their willingness to mate.[10] The very oddness and complexity of their ritual is an important part of this communication—it makes the language of the species unique and precise, reducing the possibility of making wasteful and potentially dangerous decisions. For example, it prevents the male fritillary from attempting to mate with a female of the wrong species or, for that matter, a brightly colored fluttering leaf.

This fundamental harmony of understanding between the courting butterflies is the result of what might be called a biological "resonance" between them, set up by the effect of the repetitive rhythms of the courtship flight upon their respective nervous systems. Neurobiologically, the butterflies are "vibrating" in harmony, like a pair of tuning forks. This sense of closeness and common purpose allows them to transcend the normal self-protective instincts that would usually compel them to avoid interaction with others and reap survival benefits they could not have managed on their own.[11] By transcending these instinctive restrictions, they are free to approach each other and mate.

In this simple way, ritual behaviors became rooted in the genes of primitive animals. In simple animals like the butterfly, ritual behaviors became rigid and complex—the simple nervous system of such animals could not resolve even the slightest degree of ambiguity. They must be presented with only the simplest choices: it's a butterfly or it's not; I can mate with it or I can't.

Animals with more complex neurological systems aren't so rigidly restricted—a cat, for example, has more sophisticated ways to identify another cat—and so the structure of their rituals is not as inflexible. But research shows that, even in relatively complex animals, neurological communication remains an important component of ritual behavior.[12] Repetitive, rhythmic interactions among animals—similar to those used in animal ritual—result in a high degree of limbic arousal, but only when those interactions occur among animals of the same species. The fact that this limbic activity only occurs between animals of the same species indicates that some sort of neurological resonance is occurring. The fact that it occurs at all indicates that the effects of ritual behavior can affect the most complex functions of the animal's mind, giving ritual the capability to alter emotion and mood.

In humans, because of the brain's great complexity, ritual behavior almost always involves the highest levels of thought and feeling. Our rituals are about something; they tell stories, and these stories give them meaning and power. The stories are crucial to the effectiveness of human ritual, chosen and shaped to meet specific cultural needs. But we believe that the root of the ceremonial rites of all human societies, from the most primitive to the most exalted, are an elaboration of the neurobiological need of all living things to escape the limiting boundaries of the self.[13]

THE NEUROBIOLOGY OF RITUAL

From the neurobiological perspective, human ritual has two major characteristics. First, it generates emotional discharges, in varying degrees of intensity, that represent subjective feelings of tranquillity, ecstasy, and awe; and second, it results in unitary states that, in a religious context, are often experienced as some degree of spiritual transcendence.[14] Both of these effects, we believe, are neurobiological in origin.

Ritual, Biology, and Transcendence

The ability of human ritual to produce transcendent unitary states is the result, we believe, of the effect of rhythmic ritualized behavior upon the hypothalamus and the autonomic nervous system and, eventually, the rest of the brain. Studies have shown that participating in spiritual behaviors such as prayer, religious services, meditation, and physical exertion can lower blood pressure, decrease heart rate, lower rates of respiration, reduce levels of the hormone cortisol, and create positive changes in immune system function.[15] Since all these functions are regulated by the hypothalamus and autonomic system, the effect of ritual upon autonomic states seems clear.

The process of ritual begins as rhythmic behaviors subtly alter autonomic responses in the body's quiescent and arousal systems. If those rhythms are fast—in the case of Sufi dancing, for example, or in the frenzied rites of Voudon—the arousal system is driven to higher and higher levels of activation. This increasing level of neural activity soon becomes an issue for the hippocampus, the diplomat of the limbic structure that is responsible for maintaining a sense of equilibrium in the brain.[16]

In the hippocampus, information is exchanged between various parts of the brain. Yet this center of the brain also frequently acts as a kind of floodgate, regulating the flow of neural input between various regions of the brain. This regulatory function moderates the level of neural activity, and keeps the brain in a state of relative equilibrium. For example, when the hippocampus senses that brain activity has reached excessively high levels, it exerts an inhibitory effect on neural flow—in effect, it puts the brakes on brain activity—until action in the brain settles down. As a result, certain brain structures are deprived of the normal supply of neural input on which they depend in order to perform their functions properly.

One such structure is the orientation association area—the part of the brain that helps us distinguish the self from the rest of the world and orients that self in space—which requires a constant stream of sensory information to do its job well. When that stream is interrupted, it has to work with whatever information is available. In neurological parlance, the orientation area becomes deafferented—it is forced to operate on little or no neural input. The likely result of this deafferentation is a softer, less precise definition of the boundaries of the self. This softening of the self, we believe, is responsible for the unitary experiences practitioners of ritual often describe.

The same neurobiological mechanism underlying unitary experiences can also be set in motion, in a slightly different manner, by the intense, sustained practice of slow ritual activity such as chanting or contemplative prayer. These slow rhythmic behaviors stimulate the quiescent system, which, when pushed to very high levels, directly activates the inhibitory effects of the hippocampus, with the eventual result of deafferenting the orientation area and, ultimately, of blurring the edges of the brain's sense of self, opening the door to the unitary states that are the primary goal of religious ritual.

Ritual and Emotion

The unitary experiences produced by ritual acts are almost always accompanied by strong emotional states, which are themselves a result of rhythmic behaviors.[17] Repetitive motor behaviors such as dancing or singing in ceremonies can have significant effects upon the limbic and autonomic systems, both of which are involved in creating emotion and mood. One study has shown that repetitive auditory and visual stimuli—ritualized dancing, singing, or chanting, for example—can drive cortical rhythms to produce ineffable, intensely pleasurable feelings.[18] Other work has demonstrated that rhythmic behaviors simultaneously activate several senses at once. In combination with other contributing activities that are often a part of ritual—fasting, hyperventilation, and the inhalation of incense or other fragrances—this multisensory stimulation can affect the physiology of the body in ways that can lead to altered mental states.

The emotional qualities associated with ritual-induced states appear largely the result of the effects of repetitive rhythms on the autonomic nervous system and brain. But the intensity of these emotional states can also be affected by other components of ritual behavior. For example, rituals often incorporate specific "marked actions"—a slow bow, a prostration, a deliberately excursive movement of the hands and arms, or any other action that by its form or meaning draws attention to itself as being different from ordinary, practical movements.[19] The oddness of this gesture attracts the attention of the watchdog-like amygdala. Although the amygdala is looking for signs of opportunity and danger, inexplicable movements, such as the marked actions of human ritual, might easily capture and hold its attention for longer than usual periods of time.

If stimulation of the amygdala lasts long enough, the animal often responds with fear, cringing, and withdrawal. It's possible that during ceremonial ritual, marked actions would cause a sustained focus of the amygdala, resulting in a mild fear or arousal response, just as occurred in the animals whose amygdalas were electrically stimulated. Blended with the blissful calm of the hyperquiescent state, this arousal might be experienced as "religious awe."

These feelings of awe can be further augmented by the sense of smell, which might account for the customary use of incense and other fragrances in religious rites.[20] The middle part of the amygdala receives nerve impulses from the olfactory system, so strong smells could stimulate the watchdog to generate alertness or a mild fear response. Studies have also shown that certain odors can result in very specific emotional responses: lavender, for example, evokes feelings of relaxation and calm while acetic acid has been shown to trigger feelings of anger and disgust.[21] When rituals combine fragrances with marked actions and repetitive sounds—if, for example, a priest swinging an incense dispenser pauses to make a benedictive bow during a prayer and response—the resulting stimulation to the amygdala might result in an intensification of the sense of religious awe.

The emotional power of religious ritual might also be augmented by the action of the hypothalamus, which, when stimulated by arousal activity, can trigger positive psychological states ranging from mildly pleasant sensations to feelings of ecstasy.[22]

Emotional responses of any kind are clearly related to autonomic activity.[23] This is especially true for the strong emotional feelings that are the result of ritual behaviors. But research shows that autonomic activity alone is not sufficient to create the intense emotional states experienced during ritual. For example, in several studies, researchers have found that chemical stimulation of the autonomic system does not activate the brain's emotional centers, as the rhythmic behaviors of ritual are able to do.[24] This implies that the emotional effects of ritual are dependent upon other body sensory input and, most important, the cognitive context in which religious ritual is performed.[25] It seems that more than mere autonomic stimulation is required to trigger the emotional states associated with ritual; ideas that have a deeper psychological charge or emotional pull are also required. This suggestion is supported by a study of meditation practices, in which researchers found that meditators who used a mantra that had some personal meaning (as opposed to an abstract word or sound), were more successful in reaching altered meditative states.[26]

In other words, in order for human ritual to be effective in engaging all parts of the brain and body, it must merge behaviors with ideas; it's this synthesis of rhythm and meaning that makes a ritual powerful.

In religious ritual, the deep belief in the existence of God and the human ability to communicate with this greater reality or force, gives great energy (i.e., brain stimulation) to the individual practicing prayer or chanting. Yet anthropologists have noted that even secular rituals make reference to concepts that may be considered spiritual—patriotic rituals, for example, emphasize the "sacredness" of a nation, or a cause, or even a flag, while handshakes and other greeting rituals implicitly recognize the sacredness of the individual.

In a sense, then, every ritual turns a meaningful idea into a visceral experience. The ideas that animate religious ritual are rooted in stories and myth.

THE RITUAL-MYTH CONNECTION

When the unitary states generated by the neurobiology of ritual occur in a religious context, they are usually interpreted as a personal experience of the closeness of God. In this sense, ritual offers a way to resolve neurologically the fundamental problem that all systems of mythology must address—that is, the awe-inspiring distance normally perceived between humans and their gods. According to Joseph Campbell, this existential dilemma is "the one great story of myth; that in the beginning we were united with the source, but that we were separated from it and now we must find a way to return."[27]

Restoring this original union between individuals and their spiritual source is the promise of virtually every known system of belief, from the primal myths of early hunting cultures to all the great religions that flourish today. In Christian theology, for example, Jesus provides the pathway to God; in Buddhist belief, oneness with all can be reached by following Buddha's teachings; and in Islam, reconciliation is possible through submission to the will of al-Lah.

Such assurances, spelled out in scriptures, can provide a powerful basis for faith and an effective buffer against existential fears. But these assurances are, ultimately, only ideas, and even in their most potent state, can only be believed in the mind by the mind. The neurobiology of ritual, however, turns these ideas into felt experiences, into mind-body, sensory, and cognitive events that "prove" their reality. By giving us a visceral taste of God's presence, rituals provide us with satisfying proof that the scriptural assurances are real.

Roman Catholics, for example, can experience Jesus' presence in the most intimate and unifying way, through the sacrament of the Eucharist, which fulfills Christ's promise of union and eternal life. In the same fashion, Buddhists use meditation and other contemplative rituals to transcend their attachments to the egotistical self, and all the mortal suffering those attachments cause, and to lose themselves in the serene oneness of existence that the Buddha so eloquently described. Thus, the neurobiological effects of ritualized behaviors give ceremonial substance to the stories of myth and scripture. This is the primary function of religious ritual—to turn spiritual *stories* into spiritual *experience*; to turn something in which you believe into something you can feel. It's why dervishes whirl, why monks chant, why Muslims prostrate themselves, and why primeval hunters, hoping to win the favor of the great animal spirits, donned the skins of bears and wolves, and danced reverently around the fire.

In every time and culture, it seems, humans have intuitively found ways to tap the neurobiological mechanisms that give ritual its transcendent power, by bringing their most important myths to life in the form of ritualized behaviors. "No human society has yet been found," says Joseph Campbell, "in which . . . mythological motifs have not been rehearsed in liturgies; interpreted by seers, poets, theologians, or philosophers; presented in art; magnified in song; and ecstatically experienced in life-empowering visions."[28]

What is it in the makeup of human beings that compels us to act out our myths—to "rehearse," as Campbell puts it, the unifying resolutions they so universally promise? Until recent years, anthropologists largely agreed that the urge to perform rituals was purely a cultural drive. Societies performed rituals, they believed, because humans had learned over time that ritual behaviors offered powerful social benefits. As we've mentioned, rituals do provide powerful social benefits, but our growing understanding of neurological function leads us to believe that the ritual urge may be rooted in something deeper than the cultural needs of a given society. It suggests, instead, that humans are driven to act out their myths by the basic biological operations of the brain.

Why We Act Out Myth

All our physical movements begin in a part of the brain known as the premotor area. This is the area that plans physical movements and then initiates them by instructing the brain's motor area to send the proper instructions to the appropriate groups of muscles. Precise control of movements and reactions is a crucial factor in the survival of all higher organisms, so the operations of the premotor area became an important factor in the overall functioning of animal brains.

In the human brain, the premotor area also served as the physiological base from which a related but more sophisticated structure evolved. Also called the prefrontal cortex, the attention association area is primarily responsible for focusing attention, generating a sense of will, and mediating emotion. But because of vestigial connections to the motor areas through the premotor area from which it evolved, the attention association area can also initiate movement. In fact, without the inhibiting actions performed by the brain's frontal lobe, the attention area would compel us physically to act out all our thoughts.[29]

Evidence for this inclination consists of the pathological states that can occur when these inhibitory functions are disabled by damage from stroke, tumor, or disease. The rare condition called *latah* is one such pathological state. The victims of latah are compelled physically to act out whatever command they hear and, sometimes, whatever they see. One of the first observers of this rare disorder was the nineteenth-century neurologist Guilles de la Tourette. In 1884, Tourette described his meeting, in Malaysia, with several victims of latah, whom he referred to as "jumpers."[30]

"Two jumpers who were standing near each other were told to strike," he wrote, "and they struck each other, each very forcibly. When the commands are issued in a quick, loud voice the jumper repeats the order. When told to strike, he strikes, when told to throw it, he throws it, whatever he has in his hands."

Tourette also described an encounter with a woman afflicted with the same disorder. "I talked to her for at least ten minutes, without perceiving anything abnormal in her conduct or conversation," he reported. "Suddenly, her introducer threw off his coat. To my horror, my venerable guest sprang to her feet and tore off her *kabayah*. My entreaties came too late to prevent her continuing the same course with the rest of her garments."

Tourette made it clear that these bizarre behaviors were not the result of a psychotic break with reality. Each victim, he said, was "perfectly conscious of the mental abasement which he is exhibiting, and resents his degradation."

Other similar disorders exist: victims of echopraxia, for example, compulsively repeat any action they observe, and sufferers of echolalia repeat whatever they hear.[31] Like latah, these are not psychotic states; they are neurological problems possibly caused by a breakdown in the neural inhibitory mechanisms, usually associated with the attention association area. The fact that such inhibitors are necessary in the first place to prevent us from slavishly acting out all our thoughts suggests that it is the inbuilt tendency of the brain to turn thoughts into actions. In fact, even when our brains are operating normally, this tendency may break through, as when we talk with our hands, or when we lean and sway in an attempt to coax an errant golf shot back toward the fairway.

The inborn physical compulsion to enact our thoughts may have an evolutionary purpose. By mentally rehearsing certain important actions—running, fighting, stalking, and killing prey—we might actually hone our abilities to perform those tasks in real life. Many athletes, in fact, prepare for their events by mentally visualizing themselves in action, as do musicians and other performers.[32]

If the brain contains such a compulsion to act out thoughts and ideas, it would be no surprise if the brain compelled us to act out the stories of myth. The ideas these stories convey about fate, death, and the nature of the human spirit are of immediate and lasting concern, and would most certainly gain the mind's attention. In the course of acting out mythic stories, it's possible that some individuals would stumble upon the types of rhythmic behavior that would trigger the neurological chain of events that leads to transpersonal or transcendent feelings. This palpable sense, together with the symbols and themes of the mythic story, would result in the creation—the discovery, really—of an effective ritual act.

All religious rituals, to be effective, must combine the essential content of a mythic story with the neurological reactions that bring the myth to life. The synthesis of these two elements—the sheer neurological function and the meaningful cultural content—is the true source of a ritual's power. It allows the worshiper to enter a mythic story metaphorically, confront the profound mysteries the myth embraces, and then experience the resolution of those mysteries in a powerful, possibly life-changing way. The ritual of the mass, for example, allows the individual to participate in a triumphant reenactment of Jesus' Last Supper, because the stately rhythms of the ceremony are effectively integrated with a liturgy formed from the specific teachings and promises of Christ. Each of these elements is critically important, so important, in fact, that if the ritual is to maintain meaning from one generation to the next, the balance between rhythm and content must constantly be adjusted.

In the 1960s, for example, the Roman Catholic Church, as part of a campaign of sweeping reform, translated the liturgy of the mass from the ancient Latin and Greek into popular modern languages. The obvious intention was to clarify the meaning of those ancient phrases and make the mass a more relevant experience for twentieth-century Catholics. The move was controversial at the time, and many Catholics, who'd been raised on the old Latin incantations, found the new mass much less satisfying. For those individuals, the very sound of the Latin phrases was a part of the content of the ritual. It didn't matter whether they understood the words. Their ritual had been altered in a way that had, for them, reduced its effectiveness.

But the essential content of the liturgy had been maintained and the rhythms of the ceremony were largely intact, so, in time, the ritual of the new mass became, for the majority of Roman Catholics, as effective as the old.

Virtually all rituals must maintain this delicate balance between permanence and impermanence, between the cultural content derived from the stories of myth and the neurological resonance of rhythmic behavior. If the rhythms of ritual don't generate the proper autonomic and emotional responses, the ceremony will lose its underlying power; if the symbols and themes the ritual employs grow stale or culturally irrelevant, its spiritual meaning will dissolve. In this sense, effective rituals are not only generated by the rhythms of physical or mental activity, they also require a larger, more dynamic rhythm—the slow fluctuation of stability and change—in order to stay alive.

The power of ritual lies in its ability to provide believers with experiential evidence that seems to "prove" that the guarantees made in myth and scripture are true. Rituals allow participants to taste, if only for a moment, the transcendent spiritual unity that all religions promise. The fact that the unifying effects of ritual are generated by basic biological function explains the pervasiveness of ritual activity in virtually every culture, and the similarity of purpose with which the rituals of the world have evolved. It also tells us why ritual ceremonies still have such power for so many, even in this rational age.

As we've mentioned, the spiritual intensity of the unitary states produced by religious ritual can vary tremendously. The members of a rosary group, for example, may experience a mild sense of spiritual connection, generated by the quiet cadence of their prayers. The intense physical exertion of more strenuous ritual behaviors—prolonged ceremonial dancing, for example, or the rigorous "vision quests" of various Native American tribes—can often lead to much deeper levels of unitary experience, often characterized by trance states and hyperlucid visions.

It's important to understand that all these unitary states, from the mildest to the most extreme, are triggered by the sensory effects of repetitive rhythmic behavior—that is, they begin with physical activity, and progress, in bottom-up fashion, to the mind. The more intense the physical activity and the longer it is sustained, the deeper the resulting unitary state usually will be. The intensity of those states, therefore, is finite, and is determined by the participant's physical stamina. In other words, while ritual can provide a taste of spiritual union, it is unlikely to carry us to the ultimate unitary states. The limitations of the body stand in the way.

Intriguingly, the same neurological mechanisms triggered by the physical behaviors of ritual from the bottom-up can also be triggered by the mind working in top-down fashion—that is, the mind can set this mechanism in motion, starting with nothing more substantial than a thought. The mind, of course, is not as easily tired as the body, and with practice, a single thought could be sustained indefinitely. Theoretically, the proper kind of thought would not only trigger the mechanism of transcendence, it would also push the degree of transcendence to ultimate levels, resulting in profound unitary states, the states of spiritual absorption that theologians describe as mystical experience. This is precisely what happens, we believe, when the mind is engaged in the ancient religious practices of meditation and contemplative prayer.

6

MYSTICISM

The Biology of Transcendence

To prepare herself for the holy Lenten season, a fourteenth-century German nun named Margareta Ebner spent several days absorbed in reverent silence and constant, contemplative prayer. One night, as she prayed alone in her convent's chapel, she perceived, in the choir loft, a wondrous presence, which she later described in her journal:

> Now when the hallelujah was rung at that time, I began to keep silence with the greatest joy, and especially in the night before Shrove Tuesday I was in great grace. And then it happened on Shrove Tuesday that I was alone in the choir after matins and knelt before the altar, and a great fear came upon me, and there in the fear I was surrounded by a grace beyond measure. I call the pure truth of Jesus Christ to witness for my words. I felt myself grasped by an inner divine power of God, so that my human heart was taken from me, and I speak in the truth—who is my Lord Jesus Christ—that I never again felt the like. An immeasurable sweetness was given to me, so that I felt as if my soul was separated from my body. And the sweetest of all names, the name of Jesus Christ, was given to me then with such a

> great fervor of his love, that I could pray nothing but a continuous saying that was instilled in me by the divine power of God and that I could not resist and of which I can write nothing, except to say that the name Jesus Christ was in it continually.[1]

Could it be that Sister Margareta really was visited by the mystical presence of Jesus in that lonely chapel, seven hundred years ago? Or was she, as most modern, rationalistic thinkers would insist, the victim of some emotional or psychological imbalance that the science of her time could not begin to fathom? In the common view of current scientific understanding, the ecstatic spiritual unity experienced by this God-struck German nun, and a thousand other mystics like her, wasn't spiritual at all but was, instead, a delusional state brought on by brain dysfunction or any number of psychological stresses. Medical research has proposed many causes for these intense religious states, from fatigue or emotional distress, to obsessive thinking or even mental illness. Since the time of Freud, in fact, many psychiatrists have believed that mystical experiences are illusions triggered by the neurotic, regressive urge to reject an unfulfilling reality, and recapture the bliss we knew as infants, bathed in the safe and all-encompassing unity of a mother's love.[2] The Freudian explanation of Margareta's mystical moment would sound something like this: As the sister prayed fervently in the chapel, contemplating the glories of heaven and longing to escape the emptiness of the mundane world, she somehow tapped into unconscious memories of transcendent infantile joy (possibly due to a seizure). Later, as she tried to make sense of this sudden surge of rapture, her spiritual sensibilities led her to what seemed to be the obvious explanation: She had been touched by the presence of God.

Science has no choice, of course, but to find such natural causes for "supernatural" events. From a rational point of view, it's hard to imagine that the claims of mystics could be based on anything other than delusion. Our own scientific research, however, suggests that genuine mystical encounters like Sister Margareta's are not necessarily the result of emotional distress or neurotic delusion or any pathological state at all. Instead, they may be produced by sound, healthy minds coherently reacting to perceptions that in neurobiological terms are absolutely real. The neurobiology of mystical experience makes this clear, but before we can investigate the mind's machinery of transcendence we must craft a precise definition for the term *mysticism* and understand how the insights of the mystics have helped to shape the religions of the world.

MYSTICISM DEFINED

In her book *Mysticism*, a preeminent study of mystical spirituality, author Evelyn Underhill calls the term *mysticism* "One of the most abused words in the English language.

"It has been used in different and often mutually exclusive senses by religion, poetry and philosophy," she says, "has been claimed as an excuse for every kind of occultism, for dilute transcendentalism, vapid symbolism, religious or aesthetic sentimentality, and bad metaphysics. On the other hand, it has been freely employed as a term of contempt by those who have criticized these things."[3]

In modern usage, "mysticism," like its linguistic cousin "myth," is often used pejoratively to dismiss sloppy or superstitious thinking. The *New World Dictionary*, in fact, defines the word as "vague, obscure, or confused thinking or belief." But for Underhill, there is nothing vague or confused in mystical thought. Mysticism, she says, "is not an opinion: It is not a philosophy. It has nothing in common with the pursuit of occult knowledge. . . . It is the name of that organic process which involves the perfect consummation of the Love of God: the achievement here and now of the immortal heritage of man. Or, if you like it better—for this means exactly the same thing—it is the art of establishing his conscious relation with the Absolute."

Underhill's definition is supported by the words of the mystics themselves. According to the fourteenth-century German mystic John Tauler, for example, the mystic's soul becomes "sunk and lost in the Abyss of the Deity, and loses the consciousness of all creature distinctions. All things are gathered together in one with the divine sweetness, and the man's being is so penetrated with the divine substance that he loses himself therein, as a drop of water is lost in a cask of strong wine."[4]

Mystical experience, in other words, is not about magic, or mind-reading, or the conjuring of visions or spirits; it is nothing more or less than an uplifting sense of genuine spiritual union with something larger than the self. This definition is consistently endorsed by the accounts of mystics throughout time, and in all religious traditions. Those same accounts also suggest that mystical experience is a distinct and cohesively patterned phenomenon. In 1997, neurological researchers Jeffrey Saver and John Rabin presented a paper that, in part, drew upon these accounts to define specific core elements of the mystical experience. They found that mystical states are often characterized by strong, contradictory emotions—for example, terrifying fear might coexist with overpowering joy. In mystical experience, time and space are perceived as nonexistent, and normal rational thought processes give way to more intuitive ways of understanding. The mystic frequently experiences intimations of the presence of the sacred or the holy, and often claims to have seen into the most essential meaning of things, resulting in a rapturous state that has been described as "an interior illumination of reality that results in ultimate freedom."

At the heart of all the mystic's descriptions, however, is the compelling conviction that they have risen above material existence, and have spiritually united with the absolute. The primordial longing for this absolute union, and the transcendent experiences to which it might lead, are the common threads that run through the mystical traditions of East and West, of ancient centuries, and of the present. And while the mystics of different times and traditions have used many techniques to attain this lofty union, from the pious self-denial of medieval Christian saints to the ritual sexuality of some tantric Buddhists, the mystical states they describe sound very much the same. For example, here is the Sufi master Hallaj Husain ibn Mansur, a resident of medieval Iraq, describing the intimate intermingling of the mystic and his Lord:

> I am He Whom I love, and He whom I love is I:
> We are two spirits dwelling in one body.
> If thou seest me, thou seest Him,
> And if thou seest Him, thou seest us both.[5]

The medieval Catholic sage Meister Eckhart, writing from the cooler climes of Germany, had similar words to say on the very same subject:

> How then am I to love the Godhead? Thou shalt not love him as he is: not as a God, not as a spirit, not as a Person, not as an image, but as sheer, pure One. And into this One we are to sink from nothing to nothing, so help us God.[6]

The theme of unity echoes in the Taoist wisdom of Lao-tzu . . .

> Ordinary men hate solitude.
> But the Master makes use of it,
> embracing his aloneness, realizing
> he is one with the whole universe[7]

... and in the plain-spoken insights of Black Elk, the Oglala mystic and shaman:

> Peace comes within the souls of men
> When they realize their oneness with the universe.[8]

Virtually all mystical traditions identify some sense of union with the absolute as the ultimate spiritual goal. Correspondingly, nearly all those traditions have developed rigorous systems of training and initiation, designed to help the devoted reach that rarefied state. In Zen, nonsensical koans were used to loosen the grip of the conscious mind, and open the doorway to the spirit. Kabbalistic Jews performed complicated mental manipulations of numbers and images to reach the same end. Christian mystics relied upon intense contemplative prayer, fasting, silence, and various forms of mortification to free their minds from mundane matters and focus more intently upon God. These disciplines emerged independently, but all are based on a common insight: The first step in attaining mystical union is to quiet the conscious mind and free the spirit from the limiting passions and delusions of the ego.

"The separate self dissolves in the sea of pure consciousness, infinite and immortal," says Hindu scripture.[9] "Separateness arises from identifying the Self with the body, which is made up of the elements; when this physical identification dissolves, there can be no more separate self. This is what I want to tell you, beloved."

The spiritual need to transcend the self is a central theme of Eastern religions, including Taoism, as is made clear in this excerpt from an ancient Chinese text:

> The Taoist first transcends worldly affairs, then material things, and finally even his own existence. Through this step-by-step nonattachment he achieves enlightenment and is able to see all things as One.[10]

The same ideas, however, also lie at the heart of Western schools of mysticism, and are echoed in the following words from the Hebrew mystic Rabbi Eleazar:

> Think of yourself as nothing and totally forget yourself as you pray. Only remember that you are praying for the Divine Presence. You may then enter the Universe of Thought, a state of consciousness which is beyond time. Everything in this realm is the same—life and death, land, and sea . . . but in order to enter this realm you must relinquish your ego and forget all your troubles.[11]

Thus, as Dr. Perle Epstein points out in her book on Jewish mysticism, *Kabbalah,* self-transcendence has long been an important aspiration in Jewish mysticism. In the sixteenth century, for example, Jewish mystics, following the secret teachings of the Kabbalah, aimed for the goal of *bittul hayesh* ("to annihilate the ego"). To this end, they used meditation, controlled breathing, and other contemplative techniques to silence the mind and enter a state in which they could experience, directly, the Divine Presence of God. According to Kabbalistic masters, this holy state could not be attained until the mystic had unraveled all attachments to the physical world, and obliterated any sense of an egotistical self. In the words of Rabbi Eleazar, "If you consider yourself as 'something,' and pray to Him for your needs, God cannot clothe Himself in you. God is infinite and cannot be held in any kind of vessel that has not dissolved itself into No-thing."

In similar fashion, Greek Orthodox mystics in the fifth century also came to believe that God could only be known by a mind that has been cleansed of all distracting thoughts and images. The Orthodox mystics called this stillness of mind *hesychia*, or inner silence, and taught that it was the way to open the door to a mystical union with God. In her book *A History of God*, religion scholar

Karen Armstrong explains that the goal of Greek mysticism was to gain "a freedom from distraction and multiplicity, and the loss of ego—an experience that is clearly akin to that produced by contemplatives in nontheistic religions like Buddhism. By systematically weaning their minds away from their 'passions'—such as pride, greed, sadness or anger which tied them to the ego—hesychiasts would transcend themselves and become deified like Jesus on Mt. Tabor, transfigured by the divine 'energies.' "

Armstrong finds similar ideas among the Islamic mystics, called Sufis, who developed the concept of *'fana*, or annihilation, which was brought about by a combination of fasting, sleepless vigils, chanting, and contemplation, all intended to induce altered states. These behaviors often resulted in actions that seemed bizarre and uncontrolled, which, according to Armstrong, earned those mystics who practiced such techniques the nickname of the "drunken" Sufis. The first drunken Sufi was Abu Yizad Bistami[12] who lived in the ninth century, and whose introspective disciplines, Armstrong says, carried him beyond any personalized conceptions of God.

"As he approached the core of his identity," she writes, "he felt that nothing stood between God and himself; indeed, everything that he understood as 'self' seemed to have melted away:

> I gazed upon [al-Lah] with the eye of truth and said to Him: "Who is this?" He said, "This is neither I nor other than I. There is no God but I." Then he changed me out of my identity into His Selfhood. . . . Then I communed with him with the tongue of his Face saying: "How fares it with me with Thee?" He said, "I am through Thee, there is no god but Thou."

Bistami had united with God, Armstrong says, had become a part of God, beyond his self. In Armstrong's words, "This was no external deity 'out there,' alien to mankind: God was discovered to be mysteriously identified with the inmost self. The systematic destruction of the ego led to a sense of absorption in a larger ineffable reality."

And in that mystical reality, Armstrong tells us, the rift between humans and God disappears in an inexpressibly sweet reunion.

"It would be the end of separation and sadness," she says, "a reunion with a deeper self that was also the self he or she was meant to be. God was not a separate, external reality and judge but somehow one with the ground of each person's being."

Armstrong's quote refers specifically to the mystical attainments of the Sufis, but in spirit it applies to all forms of mysticism: The goal of all mystical striving is to shed the limits of the self and return to that original condition of wholeness, the primal state of unity with God, or the cosmos, or the Absolute.

"The overcoming of all the usual barriers between the individual and the Absolute is the great mystic achievement," says William James in *Varieties of Religious Experience*.

> In mystic states we both become one with the Absolute and we become aware of our oneness. This is the everlasting and triumphant mystical tradition, hardly altered by differences of clime or creed. In Hinduism, in Neoplatonism, in Sufism, in Christian mysticism . . . we find the same recurring note, so that there is about mystical utterance an eternal unanimity which ought to make a critic stop and think, and which brings it about that the mystical classics have, as has been said, neither birthday nor native land. Perpetually telling of the unity of man with God, their speech antedates languages, and they do not grow old.[13]

James rightly understands that the essential mechanics of the mystical experience—that is, the attainment of spiritual union through detachment from the self—is rooted in something deeper and more primal than theology or scriptural revelation. But it's doubtful he would ever guess that those mechanics are wired into the human brain, and are set in motion by nothing more tangible than the mind willing itself toward God.

MYSTICISM AND MENTAL HEALTH

To modern rational sensibilities, mystics and mysticism may seem part of a dim and distant past, but in fact, a landmark study, carried out in 1975, showed that despite the astounding insights and enlightenments of the scientific age, mystical spirituality remains surprisingly prevalent in contemporary life.[14] The study, conducted by sociologist Andrew Greeley for the National Opinion Research Center, asked the question, "Have you ever felt as though you were very close to a spiritual force that seemed to lift you out of yourself?" Remarkably, more than 35 percent of the study's population answered that they had. Of that number, 18 percent reported one or two experiences, 12 percent reported "several," and 5 percent said they experienced such events "often."

In ancient and medieval cultures, mystics were often held in high esteem as the wisest and most spiritually attuned members of a society. The rationalistic and empirical demands of Western science, however, seem to leave professional observers no choice but to regard these modern mystics as the victims of damaged or deluded minds. Certainly, there is evidence to support that point of view. We know, for example, that certain pathological conditions such as schizophrenia and temporal lobe epilepsy can trigger voices, visions, and other hallucinatory effects that often possess religious connotations, and that occasionally these hallucinations can lead to an abnormal fascination with spiritual affairs.[15]

There is also the widely accepted Freudian view that regards mystical experience as the result of a regressive, infantile neurosis. Freud's theory is based upon the state of "oceanic bliss" that he believed fills an infant's mind before the lines between "self" and "not-self" have been drawn. In Freud's interpretation, the longing for spiritual union, which mystics so universally express, is really an unconscious desire to escape a harsh, disappointing reality by returning to the world of blissful unity and completeness we knew when we were babies.

Science, it seems, has no shortage of rational explanations for the strange accounts of the mystics, and while these explanations may vary in approach, they all agree on one important point: The mind of a mystic is a mind that has somehow become fundamentally confused. Mysticism, in other words, is the result of mental pathology, and mystics, whether they suffer from neurosis, psychosis, or functional problems of the brain, are people who have clearly lost track of what is real.

For many rational thinkers, even those whose reason allows room for religious belief, this conclusion seems satisfyingly scientific—and safely beyond doubt. Science, however, has not been able to empirically prove that mysticism is a product of distraught or dysfunctional minds. Significant research, in fact, seems to show that people who experience genuine mystical states enjoy much higher levels of psychological health than the public at large. For example, the Greeley study mentioned above found that respondents who claimed to have entered mystical states were in a "state of psychological well-being substantially higher than the national average" as measured by standard psychological scales. Other studies have also shown that in general, even mild mystical and spiritual experiences are associated with higher-than-average levels of overall psychological health,[16] expressed in terms of better interpersonal relationships, higher self-esteem, lower levels of anxiety, clearer self-identity, an increased concern for others, and a more positive overall outlook on life.

These findings raise an interesting question: If mystical experience is, in fact, the product of a confused or disordered mind, how can such a mind also generally demonstrate such enviable levels of mental clarity and psychological health? On the other hand, if mystical experiences are the product of healthy minds, why are the accounts of mystics so similar to the religious delusions associated with mental illness?

The answer, we believe, requires that a careful distinction be drawn between mental delusions and what we'll call "genuine" mystical states. In fact, this is exactly what researchers Saver and Rabin have done in their 1997 paper mentioned above.

In a section of the paper entitled "Delusional Disorders," the authors compare the religious hallucinations caused by schizophrenia and other psychotic states with what they refer to as "culturally accepted religious-mystical beliefs." The paper notes that there are, in fact, similarities between the two: Both are apparently delusional states defined by unusual thoughts and behaviors and a sense of separation from the normal, mundane world. Closer examination, however, reveals that these states are different in profound and specific ways.

For example, while both states may be accompanied by religious visions, voices, and other unusual events, mystics and psychotics respond to their experiences in dramatically different ways. Mystics almost always describe their experiences as ecstatic and joyful, and the spiritual unity they claim to achieve is most often described using words such as "serenity," "wholeness," "transcendence," and "love." Psychotics, on the other hand, are often confused and terribly frightened by their religious hallucinations, which are often highly distressing in nature and often include the presence of an angry, reproachful God.

Similarly, both mystics and psychotics experience what seems to be a break with normal reality. For mystics, this period of withdrawal is welcomed and even longed for. When the separation ends and they return to "normal" reality, they are able to share their experiences coherently with others, and to once again function effectively in society. For the psychotic, however, withdrawal from normal reality is an involuntary and usually distressing occurrence. Delusional psychotic states can last for years, and they inevitably drive their victims into progressively deeper states of social isolation. Mystics, on the other hand, are often among the most respected and effective members of some societies.

Finally, mystics and psychotics tend to have very different interpretations of the meaning of their experiences. Psychotics in delusional states often have feelings of religious grandiosity and inflated egotistical importance—they may see themselves, for example, as special emissaries from God, blessed with an important message for the world, or with the spiritual power to heal. Mystical states, on the other hand, usually involve a loss of pride and ego, a quieting of the mind, and an emptying of the self—all of which is required before the mystic can become a suitable vessel for God.

These distinct differences make a very strong case that mysticism is not a product of psychotic delusion, but religious hallucinations are also associated with pathological conditions other than psychosis. For example, some types of temporal lobe epilepsy can trigger spontaneous hallucinatory events that strongly resemble the experiences described by mystics. According to Saver and Rabin, the effects of epileptic seizures upon the temporal lobe of the brain have been associated with sensations of sudden ecstasy and religious awe; with increased interest in religion and even religious conversions; with out-of-body experiences; with the apprehension of the "unity, harmony, joy, and/or divinity of all reality," and in some cases, with the perceived presence of God. They describe, for example, an older woman whose seizures were characterized by visions of God and the sun. "My mind, my whole being," she said, "was pervaded by a sense of delight." They also refer to a patient "whose seizures consisted of feelings of detachment, ineffable contentment, and fulfillment; visualizing a bright light recognized as the source of knowledge; and sometimes visualizing a bearded young man resembling Jesus Christ."

Many researchers have found the link between epilepsy and spirituality very compelling. Some researchers have even gone so far as to posthumously diagnose history's greatest mystics as victims of epiletic seizures. Some of these diagnoses suggest, for example, that Mohammed, who heard voices, saw visions, and sweated profusely during his mystical interludes, may have suffered from complex partial seizure. The same type of seizure may have been the source of the blinding light that struck St. Paul on the road to Damascus and caused the auditory hallucinations that led him to believe he had heard the voice of Jesus. Joan of Arc, who also saw a spiritual light and was transfixed by beatific voices, may have suffered ecstatic partial seizures and perhaps an intracranial tuberculoma. Various epileptic states may also have been responsible for the visions of the Catholic mystic Saint Teresa of Avila, the conversion experience of Mormon patriarch Joseph Smith, the ecstatic trance states of Emanuel Swedenborg, even the hyper-religiosity of Vincent Van Gogh.

These are intriguing speculations, and there are clear overlaps between some of the symptoms of epileptic seizures and some aspects of mystical behavior. Still, we do not believe that genuine mystical experiences can be explained away as the results of epileptic hallucinations or, for that matter, as the product of other spontaneous hallucinatory states triggered by drugs, illness, physical exhaustion, emotional stress, or sensory deprivation.[17] Hallucinations, no matter what their source, are simply not capable of providing the mind with an experience as convincing as that of mystical spirituality.

We base this opinion on some very simple observations. In the case of hallucinations caused by seizure, we know that until the underlying cause of seizure activity is resolved, seizures tend to strike frequently and with regularity. In serious cases, a seizure victim may experience several attacks per week, or even per day. Most mystics, on the other hand, experience only a handful of mystical encounters in a lifetime. The hallucinatory states caused by seizures also tend to be consistent and repetitive in pattern—the victim hears the same voice with the same message, for example, or feels the advent of the same inexplicable rapture. But the spiritual episodes reported by mystics are as variable as ordinary experience: the emotional tone might differ from time to time, an angelic voice might have a different message, and so on.

Mystical experiences are also set apart, from all hallucinatory states, by the high degree of sensory complexity they usually involve. First, hallucinations usually involve only a single sensory system—a person may *see* a vision, *hear* a disembodied voice, or *feel* a sense of presence, but rarely are multiple senses simultaneously involved. Mystical experiences, on the other hand, tend to be rich, coherent, and deeply dimensioned sensory experiences. They are perceived with the same, and in some cases increased, degree of sensory complexity with which we experience "ordinary" states of mind. In plainest terms, they simply *feel* very real.

Hallucinations, of course, also feel real while they persist, but when hallucinating individuals return to normal consciousness, they immediately recognize the fragmented and dreamlike nature of their hallucinatory interlude, and understand that it was all a mistake of the mind. Mystics, however, can never be persuaded that their experiences were not real. This sense of realness does not fade as they emerge from their mystical states, and it does not dissipate over time.

"God visits the soul in a way that prevents it doubting when it comes to itself that it has been in God and God in it," says Teresa of Avila, "and so firmly is it convinced of this truth that, though years may pass before this state recurs, the soul can never forget it, to doubt its reality. . . ."[18]

In other words, the mind remembers mystical experience with the same degree of clarity and sense of reality that it bestows upon memories of "real" past events. The same cannot be said of hallucinations, delusions, or dreams. We believe this sense of realness strongly suggests that the accounts of the mystics are not indications of minds in disarray, but are the proper, predictable neurological result of a stable, coherent mind willing itself toward a higher spiritual plane.

THE NEUROBIOLOGY OF MYSTICAL EXPERIENCE

It's difficult for those of us living in the practical world, and coping from day to day with the mundane challenges of existence, to comprehend in any meaningful way the mystical sense of transcendent unity described by saints and sages. It all seems so hopelessly cryptic, so unlikely, so irrelevant to what we think of as real. In its essence, however, mystical experience is not as strange as it seems, and the first step in understanding its nature is to realize that it happens to all of us, all of the time.

Humans, in fact, are natural mystics blessed with an inborn genius for effortless self-transcendence. If you ever "lost yourself" in a beautiful piece of music, for example, or felt "swept away" by a rousing patriotic speech, you have tasted in a small but revealing way the essence of mystical union. If you have fallen in love or have ever been wonder-struck by the beauty of nature, you know how it feels when the ego slips away and for a dazzling moment or two you vividly understand that you are a part of something larger.

Like all experiences, moods, and perceptions, these unitary states are made possible by neurological function. More specifically, they are the result of the softening of the sense of self and the absorption of the self into some larger sense of reality that we believe occurs when the brain's orientation area is deafferented, or deprived of neural input.

We've already seen how the rhythmic behaviors of religious ritual can set the mechanism of deafferentation in motion, and how that process can lead to moments of transcendent spiritual unity. The same chain of events can be set in motion less formally by patterns of behavior that have no spiritual intent but are, nonetheless, ritualistic.

For example, imagine that you have just come home from a hard day at work. It is Friday night and a pleasant weekend lies ahead. To soak away the cares of the workweek, you decide to take a leisurely bath. You light a few candles, pour yourself a glass of wine, tune the radio to your favorite station, and slip into the tub.

Without intending to, you have nicely set the stage for ritual. The candles, the wine, the relaxing effects of the bath have all helped mark this moment with a sense of occasion. Like the atmospheric accents and marked actions that enhance the effects of religious ritual, these elements inform the mind, by stimulating limbic and autonomic activity, that something special is going on.

As you relax in the tub, the radio begins to play a soft romantic ballad. The slow, steady rhythms activate the body's quiescent system. As quiescent activity rises, it causes the hippocampus to exert a mild inhibition on neural flow, which would cause a slight deafferentation of the orientation area, and a mild unitary state that might be experienced as a very pleasant surge of serenity.

If the music continues, however, and quiescent levels continue to rise, that serenity might deepen into something more intense, as prolonged activation of the calming response causes the orientation area to become more effectively blocked. This more extensive blockage would result in a stronger unitary state that might feel to you as if you were being blissfully absorbed into the music.

In this manner, a quiet song, in the proper setting, can lead to a state of mind-altering self-transcendence similar to the unitary states produced by ritual. The same effect can be achieved through other mood-shifting rhythmic behaviors. Slow rhythmic activities such as reading a poem, rocking a baby, or praying can have one type of effect while fast rhythmic activities such as distance running, having sex, or cheering along with a crowd of thousands at a football game can have another. However, both fast and slow rituals can drive the brain to unitary states even though they may do so through slightly different mechanisms.[19]

In each case, rhythmic behaviors can lead to unitary states by causing the orientation area to be blocked from neural flow. The intensity of those unitary states depends upon the degree of neural blockage. Since the degree of that blockage can increase by any increment, and theoretically until there is a total blocking, a large spectrum of increasingly unitary states is possible. We call this span the *unitary continuum*. The arc of this continuum links the most profound experiences of the mystics with the smaller transcendent moments most of us experience every day, and shows that, in neurological terms, the two are different essentially by degree.

The most familiar point along the continuum is the baseline state of mind in which we experience most of our daily lives. We eat, we sleep, we work, we interact with others, and while we are consciously aware that we are, in some fashion, connected to the world around us (as part of a family, neighborhood, nation, and so on), we experience that world as something from which we are clearly set apart.

As we move up the unitary continuum, however, that separation becomes less and less distinct. We might be moved to a state of mild unitary absorption by art, or music, or walks in the autumn woods. We may reach deeper unitary states during periods of intense concentration or through the transforming intoxication of romantic love.

These activities, and the transcendent states they produce, are not religious in any formal sense, but in neurological terms they are similar to many unitary experiences produced by religious activity. These religious experiences exist along the same neurological continuum, and like all nonspiritual unitary states, their intensity is determined by the degree to which the orientation area is blocked from neural flow.

At low levels, this blockage results in mild unitary sensations, such as the feelings of unity and common inspiration shared by worshipers in a moving religious service. As we move along the continuum we find a progression of increasingly intense unitary states, characterized by feelings of spiritual awe and rapture. Where prolonged and rigorous rituals are involved, trance states may occur, featuring moments of ecstasy and hyperlucid visions. And at the farthest end of the continuum, where deafferentation would be most advanced, we find the profound states of spiritual union that have been described for us by the mystics.

As we've already explained, these advanced unitary states are usually beyond the reach of physically based ritual. The human body generally cannot sustain sufficient ritual intensity for a long enough period of time to allow deafferentation to proceed to such extremes. Mystics of all traditions have intuitively understood this, and have taught themselves to focus the inexhaustible meditative powers of the mind to carry them to these deeper states of union with the divine.

In various mystical traditions, meditative techniques assume different forms and functions. Some mystics meditate to bring the mind to a laser-sharp focus; others meditate to undo the mind's focus and sweep away all thoughts. Some engage in reflective prayer—contemplating sacred mysteries or certain lines of scripture. Others pray more passively, by simply opening their minds to the possibility of God. And others still use various combinations of these techniques to further their spiritual quests. But no matter what specific methods any given tradition of mysticism might employ, the purpose of these methods is almost always the same: to silence the conscious mind and free the mind's awareness from the limiting grip of the ego. What fascinates us, as scientific observers, is that virtually all these mystical techniques seem to have been intuitively devised to trigger the process of deafferentation, in addition to other related brain functions, and push it far beyond the levels made possible by ritual.

In broad terms, meditative techniques fall into two general categories. There are *passive* approaches, in which the intention is to clear the mind of all conscious thought, and *active* approaches, in which the goal is to focus the mind completely on some object of attention—a mantra, for example, or some symbol or scriptural verse.[20] To illustrate the ways in which meditation can trigger the neurological process of mystical self-transcendence, and lead to powerful unitary states, we'll begin with a model of brain function, based on the neurobiology we have presented, associated with the passive style of meditation, in its most basic form.

The Passive Approach

All mystical spirituality begins as an act of will. In our model, passive meditation, which is practiced in various forms by many Buddhist orders, begins with the willful intention to clear all thoughts, emotions, and perceptions from the mind. This conscious intention is instated by the brain's right attention association area—the primary source of willed actions—as the need to shield the mind from the intrusion of sensory, as well as cognitive, input.[21] To this end, the attention area, via the thalamus, causes the limbic structure known as the hippocampus, an important center of information exchange between various parts of the brain, to dampen the flow of neural input. This neural blockage affects many brain structures, including the orientation association area, which becomes increasingly deprived of information (deafferented).[22]

In the initial moments of meditation, deafferentation is only slight, but we believe that as the meditative state deepens, and the attention area tries more intensely to keep the mind clear of thoughts, this area, in conjunction with the hippocampus, chokes off more and more neural flow. As this blockage continues, bursts of neural impulses begin to travel, with increasing energy, from the deafferented orientation area, down through the limbic system, to the ancient neural structure known as the hypothalamus. The hypothalamus links higher brain activity with the basic functions of the autonomic nervous system and controls the autonomic system's ability to create both calming and arousal sensations.[23]

The impulses now reaching the hypothalamus have a powerful effect on the section of that organ structure for creating strong quiescent sensations. This causes the discharge of a burst of neural impulses back up through the limbic system and ultimately back to the attention association area. The attention area registers these calming impulses and relays them back down the circuit for one more lap. In this fashion, a reverberating circuit is established in the brain, with a stream of neural impulses gathering strength and resonance as they race again and again along their neural speedway, fostering deeper and deeper levels of meditative calm with every pass.

Meanwhile, the meditator's continued intention to clear his mind of thoughts causes a progressive buildup of neural energy, which prompts an even more aggressive effort to choke off the flow of sensory input to the orientation area. This results in higher and higher levels of deafferentation, and an ever-increasing rate of neural discharge down through the limbic system to the hypothalamus. This continued neural bombardment soon pushes the hypothalamic calming function to its limits.

Usually, such high levels of quiescent activity would cause a corresponding decrease in arousal function. Under certain conditions as we have described, however, a neurological "spillover" can occur in which the maximal activation of the calming system triggers an instantaneous maximal arousal response.

As the quiescent and arousal systems both surge, the mind is overwhelmed by simultaneous floods of calming and arousal responses. This results in an explosion of frantic neural activity, flashing up from the hypothalamus through the limbic system and back to the attention association area, which is forced, by the sudden surge, to operate at its own maximal rates. In response, the deafferenting effect that the attention area is directing toward the orientation area becomes supercharged, and in milliseconds, the deafferentation of the orientation area becomes complete.

The total shutdown of neural input would have a dramatic effect on both the right and left orientation areas. The right orientation area, which is responsible for creating the neurological matrix we experience as physical space, would lack the information it needs to create the spatial context in which the self can be oriented. Its only option, when totally deprived of sensory input, would be to generate a subjective sense of absolute spacelessness, which might be interpreted by the mind as a sense of infinite space and eternity; or conversely, as a timeless and spaceless void.

Meanwhile, the left orientation area, which we have described as crucial in the generation of the subjective sense of a self, would not be able to find the boundaries of the body. The mind's perception of the self now becomes limitless; in fact, there is no longer any sense of self at all.

In this state of total deafferentation of the orientation area, the mind would perceive a neurological reality consistent with many mystical descriptions of the ultimate spiritual union: There would be no discrete objects or beings, no sense of space or the passage of time, no line between the self and the rest of the universe. In fact, there would be no subjective self at all; there would only be an absolute sense of unity—without thought, without words, and without sensation. The mind would exist without ego in a state of pure, undifferentiated awareness. The name Gene and I have used for this state of pure mind, of an awareness beyond object and subject, is Absolute Unitary Being, the ultimate unitary state.

The mystical traditions of the East have all described some version of this ineffable unity—Void Consciousness, Nirvana, Brahman-atman, the Tao—and all hold it up as the essence of what is inexpressibly real. On the neurological level, these states can be explained as a sequence of neural processes set in motion by the willful intention to quiet the conscious mind, which is the age-old goal of passive meditation.

In a similar sense, active meditation—which consists of intensely focused contemplation or prayer—triggers a slightly different pattern of brain activity which may account for Western conceptions of the transcendent absolute.

The Active Approach

Active types of meditation begin not with the intention to clear the mind of thoughts, but instead, to focus it intensely upon some thought or object of attention. A Buddhist might chant a mantra, or focus upon a glowing candle or a small bowl of water, for example, while a Christian might pray with the mind trained upon God, or a saint, or the symbol of a cross.

For the sake of discussion, let's imagine that the focus of attention is the mental image of Christ. The process begins, just as in the passive approach, with the attention association area translating, into neurological terms, the conscious intention to pray. But in this case, since the intention is to focus more intensely upon some specific object or thought, the attention area facilitates, rather than inhibits, neural flow. In our model, this increased neural flow causes the right orientation area, in conjunction with the visual association area, to fix the object of focus, real or imagined, in the mind. Continuous fixation upon this image, induced by sustained contemplation, causes discharges from the right attention area to travel down through the limbic system to the hypothalamus, triggering the arousal section of that structure, resulting in a mildly pleasant state of excitation. As contemplation deepens, the flow of these discharges increases in intensity, until the arousal function of the hypothalamus reaches maximal levels. At this point, spillover occurs, causing the immediate maximal activation of the hypothalamus's quiescent function.

The simultaneous activation of both arousal and quiescent functions sends a flood of maximal stimulation surging back up through the limbic structures to both sides of the attention association area. As a result of this sudden neural flood, activity in the attention area is pushed to maximal levels, which amplifies the mind's ability to focus upon the object of attention, causing significant repercussions in both the left and right orientation areas.

In the left orientation area, we observe the same result we saw in passive meditation—the restriction of neural flow exerted by the hippocampus leads to deafferentation, and a consequent blurring of the sense of self. The effect on the right orientation area, however, is quite different. Remember that the attention association area has been driving the right orientation area to focus more and more intensely upon the image of Jesus. Now, as the attention area reaches maximal levels, it does not block the flow of information to the right orientation area, as it does to the left; on the contrary, the attention association area drives the right side to focus more and more intensely upon the image of Christ.

To bring the mind's focus more sharply upon this image, the attention area also begins to deprive the right orientation area of all neural input not originating from the contemplation of Jesus. In other words, the right orientation area, as it strives to create the spatial matrix in which the self can exist, has nothing to work with but the input streaming in from the attention area. It has no choice, therefore, but to create a spatial reality out of nothing but the attention area's single-minded contemplation of Christ. As the process continues, as all irrelevant neural input is stripped away and the mind becomes more focused, the image of Jesus "enlarges" until it becomes perceived by the mind as the whole depth and breadth of reality.

As these changes unfold in the right orientation area, the deafferentation of the left orientation area is also in progress, causing the perceived limits of the self to soften. As the sense of self is completely undone, by complete deafferentation, the mind would experience a startling perception that the individual self had been mystically absorbed into the transcendent reality of Jesus.

In this fashion, neurology could explain the *Unio Mystica*—the Mysterious Union with God that characterizes the spiritual experiences of so many Christian mystics, Sister Margareta included. It would, in fact, provide a neurological explanation for any mystical encounter in which the presence of a personal deity is perceived.

The experience of the *Unio Mystica* would be inexpressibly profound. It's important to understand, however, that this state of mystical union is not the same as the ultimate transcendent state, Absolute Unitary Being, in which no sense of self is possible, and no specific images of God or even of reality can exist. It's likely, though, that if active meditation carries a mystic as far as the *Unio Mystica*, it may carry him or her even further, to the ultimate unitary state. This would occur as the mystic tires, and the willed intentions of the attention association area weakens. The mind would relax its efforts to focus concentration, depriving the right orientation area of its only neural input, and sending it, along with the left orientation area, into a state of total deafferentation. At this point, the mind would enter the same selfless and limitless reality that can be reached through the act of passive meditation, the reality of absolute unitary being.

Neurologically, and philosophically, there cannot be two versions of this absolute unitary state. It may *look* different, in retrospect, according to cultural beliefs and personal interpretations—a Catholic nun, for whom God is the ultimate reality, might interpret any mystical experience as a melting into Christ, while a Buddhist, who does not believe in a personalized God, might interpret mystical union as a melting into nothingness. What's important to understand, is that these differing interpretations are unavoidably distorted by after-the-fact subjectivity. While in the state of absolute unitary being, subjective observations are impossible; on the one hand, no subjective self exists to make them, and on the other, there is nothing distinct to be observed; the observer and the observation are one and the same, there are no degrees of difference, there is no *this* and no *that*, as the mystics would say. There is only absolute unity, and there cannot be two versions of any unity that is absolute.

ABSOLUTE UNITARY BEING, EVOLUTION, AND SELF

The achievement of Absolute Unitary Being is a rare event, but even those who fall far short find themselves in spiritual states of inexpressible power and sublimity. We believe that all mystical experiences, from the mildest to the most intense, have their biological roots in the mind's machinery of transcendence. To say this in a slightly more provocative way, if the brain were not assembled as it is, we would not be able to experience a higher reality, even if it did exist.

This forces us to address a difficult but fascinating question: Why would the human brain, which evolved for the very pragmatic purpose of helping us survive, possess such an apparently impractical talent? What evolutionary advantage would a mystically gifted mind provide?

We can only speculate of course, but the nature of the evolutionary process suggests that the mind's ability to enter unitary states did not evolve specifically for the purpose of spiritual transcendence. Evolution is pragmatically short-sighted; it favors adaptations that provide effective survival advantages in the practical here and now. Those adaptations that increase an organism's chances of survival are genetically passed along; those that don't are ruthlessly winnowed out.

It's difficult to imagine what survival advantages the neurology of transcendence would offer in its partially evolved, developmental stages, and hard to find a reason why natural selection would tolerate these neural developments, which wouldn't be operational for untold millions of years.

Evolution, after all, doesn't plan ahead. It gropes impartially for potential, with no idea where that potential might lead. Consider, for example, the evolution of avian flight. No bird can fly without a fully developed set of wings, but the evolution of those wings required countless generations, with long intermediate stages during which the slowly changing wings were incapable of flight. Those intermediate stages pose a problem, which biologist Stephen Jay Gould has noted. "Of what possible use are the imperfect incipient stages of useful structures?" he asked in 1977 in *Natural History* magazine. "What good is half a jaw or half a wing?"

In other words, evolution would not equip generations of species with partially formed wings in a forward-looking campaign to produce the phenomenon of flight. It's more likely that some species began to develop small winglike appendages which served some useful purpose—perhaps, for example, they helped the animal to more efficiently dissipate heat. If these appendages worked well, they would evolve into larger structures. Eventually, they might be large enough to enable an animal to glide. Although gliding had not been the evolutionary goal of the appendages, it soon proved to have survival advantages, and at that point, the evolution of flight began.

It's likely that many of the human mind's higher functions—

our ability to ponder philosophical concepts, or to experience complex emotions such as love, grief, and envy—developed in a similar fashion from simpler neurological processes that evolved to address more basic survival needs.

We believe, in fact, that the neurological machinery of transcendence may have arisen from the neural circuitry that evolved for mating and sexual experience. The language of mysticism hints at this connection: Mystics of all times and cultures have used the same expressive terms to describe their ineffable experiences: *bliss*, *rapture*, *ecstasy*, and *exaltation*. They speak of losing themselves in a sublime sense of union, of melting into elation, and of the total satisfaction of desires.

We believe it is no coincidence that this is also the language of sexual pleasure. Nor is it surprising, because the very neurological structures and pathways involved in transcedent experience—including the arousal, quiescent, and limbic systems—evolved primarily to link sexual climax to the powerful sensations of orgasm.

The mechanism of orgasm is activated by repetitive, rhythmic stimulation. Significantly, orgasm requires the simultaneous stimulation of both the arousal and quiescent systems.[24] As we have seen, the simultaneous activation of those two systems are intimately involved in the process that sets in motion the mind's machinery of transcendence; mystical union and sexual bliss, therefore, use similar neural pathways. This does not mean, however, that they are the same experience. Neurologically, in fact, the two are quite different.

Sexual bliss is primarily generated by the hypothalamus, a relatively primitive structure, and while higher thought processes might be involved in enhancing the pleasures of intercourse, the ecstasies of sex are primarily the result of physical, tactile sensations. Transcendent experiences, on the other hand, at least those that lead to the intense unitary states described in this chapter, likely depend upon the involvement of higher cognitive structures, especially those in the frontal lobe and other association areas. The subtle and complex mental rhythms of meditation and contemplative prayer are the neurological triggers that set the process in motion.

An evolutionary perspective suggests that the neurobiology of mystical experience arose, at least in part, from the mechanism of the sexual response. In a sense then, mystical experience may be an accidental by-product, but this does not necessarily diminish the meaning of spiritual experience. Many of the brain's most sublime and sophisticated functions evolved from humbler neurological processes. The complex cognitive abilities that enable us to appreciate music or create works of art, for example, developed from simpler neural structures that evolved to address mundane survival needs. And the fact that the eagle's power of flight evolved accidentally from some thoroughly earthbound trait does not diminish the beauty, or the "truth," of flight. The potential for flight always existed. When evolution finally stumbled in the right direction, that potential was realized and the wonder of flight became real.

What is important to remember is that no matter how unlikely or unfathomable the accounts of the mystics may seem, they are based not on delusional ideas but on experiences that are neurologically real. Our understanding of brain processes predicts this, and our SPECT scan studies of Buddhists and Catholic nuns show that meditative, contemplative minds strive in that direction.

The neurological realness of Absolute Unitary Being is by no means proof of an absolute spiritual reality (we'll examine the spiritual ramifications of Absolute Unitary Being in a later chapter). On the other hand, by explaining mystical experience as a neurological function, we do not intend to suggest that it can't be something more. What we do suggest is that scientific research supports the possibility that a mind can exist without ego, that awareness can exist without self. In the neurological substance of Absolute Unitary Being, we find rational support for these inherently spiritual concepts, and a scientific platform from which to explore the deepest implications of mystical spirituality.

This exploration will ultimately lead us to some surprising conclusions about the true nature of human spirituality, and raise some fascinating questions about the processes through which the mind and the brain determine what is real. But first we will discuss the link between mysticism and religion, and examine the ways in which neurobiological functions have influenced the various incarnations of God.

7

THE ORIGINS OF RELIGION

The Persistence of a Good Idea

God is dead
. . . Nietzsche
Nietzsche is dead
. . . God

—Graffito

When philosopher Friedrich Nietzsche, in 1885, made his famous proclamation that God was dead, he was saying, of course, that God had never really lived at all. Like other great rationalistic thinkers of the nineteenth and early-twentieth centuries—Marx, Freud, James Frazer, Ludwig Feuerbach, and Bertrand Russell, to name a few—Nietzsche regarded God as just another vestige of an unscientific past that humanity would soon outgrow. It was the great expectation of many in that world-changing generation of thinkers that, as educational levels rose and science provided more realistic explanations for the mysteries of existence, the irrational appeal of religion would simply fade, and God, in all his incarnations, would simply go away.

God, however, has not obliged, and as we enter the new millennium—an age of unprecedented scientific and technological enlightenment—religion and spirituality continue to thrive. If Nietzsche and his contemporaries were alive to see it, they would most likely regard God's survival as a triumph of ignorance over reason. Convinced that religious belief is based on superstition and fearful self-delusion, they would have no choice but to conclude—as many modern rationalists have done—that humans cling to God because they lack the strength and courage to face the world without Him.

This cynical interpretation is so firmly entrenched in the thinking of rational materialism that few rationalistic thinkers would even bother to question it, but an open-minded inquiry shows that this idea may not be as intellectually sound as it seems. We believe, in fact, that the remarkable tenacity of religion is rooted in something deeper, simpler, and healthier than weak-minded denial or sheer psychological dependence.

Evidence suggests that the deepest origins of religion are based in mystical experience, and that religions persist because the wiring of the human brain continues to provide believers with a range of unitary experiences that are often interpreted as assurances that God exists. As we have seen, it's unlikely that the neurological machinery of transcendence evolved specifically for spiritual reasons. Still, we believe that evolution has adopted this machinery, and has favored the religious capabilities of the religious brain because religious beliefs and behaviors turn out to be good for us in profound and pragmatic ways.

A considerable body of research tells us this is true. Studies have shown that men and women who practice any mainstream faith live longer, have fewer strokes, less heart disease, better immune system function, and lower blood pressure than the population at large.[1] So impressive are the health benefits of religion, in fact, that after reviewing more than a thousand studies on the impact of religion upon health, Dr. Harold Koenig of Duke University Medical Center recently told *The New Republic*, that "Lack of religious involvement has an effect on mortality that is equivalent to forty years of smoking one pack of cigarettes per day."

Religion, it seems, is at least as good for the body as it may be for the soul, but the health benefits of religious behaviors do not end with physiology; a growing body of research is making it clear that religion can also be linked to superior mental health. This idea comes as a surprise to much of the modern psychiatric community that, still following in the footsteps of Freud, has long regarded religious behavior at best, as a dependent state and at worst, a pathological condition. Until 1994, for example, the American Psychiatric Association officially classified "strong religious belief" as a mental disorder.[2]

New data, however, indicates that religious beliefs and practices can improve mental and emotional health in several significant ways.[3] For example, research shows that rates of drug abuse, alcoholism, divorce, and suicide are much lower among religious individuals than among the population at large. It also seems clear that people who practice religion are much less likely to suffer from depression and anxiety than the population at large, and that they recover more quickly when they do. Other experiments have linked specific religious activities to positive psychological results; spiritual practices such as meditation, prayer, or participation in devotional services, have been shown to reduce feelings of anxiety and depression significantly, boost self-esteem, improve the quality of interpersonal relationships, and generate a more positive outlook on life.

Research has produced no conclusive reason for the healthy effects of religion, but it's a good guess that the behaviors and attitudes fostered by religions play an important role. By frowning on promiscuous sex, drugs, alcohol abuse, and other risky indulgences, for instance, and by encouraging lifestyles of moderation and domestic stability, most religions automatically promote behaviors that are inherently healthy.

The strong social support networks that characterize religious communities are almost certainly another beneficial factor.[4] The emotional support of friends and family members is an obvious and important element in anyone's mental health, but strong communities can exert a positive and very practical effect on physical well-being, too. These benefits are especially important for the elderly who, in close-knit communities, are less likely to fall into isolation, and more likely to have help with meals, medications, transportation to doctors appointments and other day-to-day activities which contribute to better physical health.

Religious behaviors may also contribute more directly to good health through the effects they exert upon the body's arousal and quiescent systems. A quiet prayer, a stately hymn, or an hour spent in meditation, can activate the body's quiescent function that has been shown to enhance immune system function, lower heart rates and blood pressure, restrict the release of harmful stress hormones into the blood, and generate feelings of calmness and well-being.[5] Obviously, any behaviors that consistently produce this quiescent response, would promote higher levels of both mental and physical health.

RELIGION AND CONTROL

It seems clear that the physical and mental benefits of faith can be linked to the values that religions promote.[6] Perhaps the healthiest aspect of religion is its power to alleviate existential stress by granting us a sense of control over an uncertain and terrifying world. This control is believed to be made possible by the presence of a formidable higher power—some knowable god, spirit, or unchanging absolute—that is willing and able to intervene on our behalf. In some cases, this higher power takes the shape of a powerful spiritual being that can be persuaded, through prayer, sacrifice, or other religious means, to champion our interests and defend us from harm. In other incarnations—in Buddhism for example—this divine personal ally becomes an impersonal refuge of ultimate spiritual wholeness and truth, in which believers can find transcendent release from the sufferings of life.

Faith in a higher power offers believers the assurance that their lives have meaning and purpose, that they are not alone in the struggle for survival, that powerful, benevolent forces are at work in the world, and that despite the terrors and uncertainties of existence, they should not be afraid.

This ability to alleviate existential gloom and connect us with powerful spiritual forces is the greatest earthly gift religions have to offer. For the individual believer, it is a gift of hope and solace, but on a larger scale, its effects have been more profound. By lifting us out of fear and futility, and giving us the sense that wise and capable hands are steering the cosmic bus, religion has served as a powerful source of confidence and motivation that has not only shaped much of human history, but may also have been a crucial reason the human race has managed to survive.

Imagine, for example, the challenges facing our Neanderthal cousins, who, as the Ice Age reigned, were still locked in fierce competition with powerful animal rivals for the mastery of the earth. In most cases, the animals were blessed with superior strength and speed, and far keener senses. Survival was always in doubt, and all that allowed our ancestors to compete was their unprecedented intelligence, and their energetic will to survive.

Unlike the animals, of course, early humans were burdened by the awareness of the certainty of death, an enervating insight that could have sent them spiraling into depression and apathy, leading to an early and unhappy end to the human saga. After all, the last thing our ancestors needed, in the ruthless tournament of natural selection, was to be taken off their game by the soul-sapping notion that no matter how hard they struggled, how skillfully they hunted, how fiercely they battled, or how creatively they thought, death was always waiting, and that their lives added up to nothing in the end.

The promises of religion protected early humans from such self-defeating fatalism, and allowed them to struggle tirelessly but optimistically for survival. Archaeological evidence tells us that even the earliest human cultures practiced rudimentary religions. Conventional thinking among many psychologists and sociologists explains the rise of religion as a cognitive process, based on faulty logic and incorrect deductions: In very simple terms, we feel fear and we long for comfort so we dream up a powerful protector in the sky.

A neurological approach, however, suggests that God is not the product of a cognitive, deductive process, but was instead "discovered" in a mystical or spiritual encounter made known to human consciousness through the transcendent machinery of the mind. In other words, humans do not cognitively invent a powerful God and then depend upon this invention to gain the feeling of control; instead, God, in the broadest and most fundamental definition of the term, is *experienced* in mystical spirituality. These intimate, unitary experiences of the presence of God makes the possibility of control apparent.

THE ORIGINS OF RELIGION

In chapter six, we saw how prolonged meditation can trigger rhythms in the mind that might lead to a neurologically based experience of spiritual union. But mystical experiences do not come only to those who strive to have them, they also come spontaneously to people who do not desire them, perhaps have never heard of them, and at first, do not know what to make of them. These spontaneous mystical occurrences range from subtle but life-changing illuminations to sudden, inexplicable raptures, from the inexpressible apprehension of an unseen spiritual presence, to the total immersion of the self into the essence of the ultimate or the divine. In a sense, all mystical experiences are spontaneous—even the mystics who dedicate their lives to the pursuit of spiritual union can't anticipate when it will occur. We believe, however, that the neurological machinery of transcendence can also be set in motion by patterns of thought and behavior not intentionally designed to provoke unitary states. These unintentional unitary experiences would provide a potent foundation for the development of religious faith; in most cases, in fact, they would make the development of religion inevitable.

Imagine, for example, a prehistoric deer hunter whose clan is in the midst of famine. Desperate for food, he hunts continuously, forsaking sleep and spending long hours alone in the wilderness. Even when he rests, he anxiously scans the horizon for signs of game, picturing in his mind the image of a magnificent stag, an animal large enough to feed the entire clan, and save all his family and friends from starvation.

Days pass, and as the hunter grows weak from hunger and fatigue, the image of the stag becomes more and more vivid in his imagination. He sees it grazing beyond the crest of a hill or drinking at the banks of a winding river. The vision soon consumes him, and his longing for a kill becomes a kind of mantra. His thoughts become repetitive, his mental focus grows more narrow and intense. Soon, his mind has been swept clear of all irrelevancies, there is no room in his consciousness for anything but the longing for the stag.

The hunter's mental focus has no spiritual component; his intention is simply to survive. From a neurological perspective, however, he is setting in motion the same biological chain of events triggered by the contemplative techniques of religious mystics as they strive to clear their awareness of any thoughts other than God.

In chapter six, we saw how "active" meditation, or contemplative prayer, could create a unitary state in which the object or image of contemplation takes on divine dimensions and the self experiences the presence of what feels like ultimate truth. It's possible that the hunter's preoccupation with the image of the stag could trigger the same neurological reaction and lead him to a similar unitary state. Just as medieval mystics might feel joyfully absorbed into the transcendent reality of Jesus, and Sufis might experience the tangible presence of al-Lah, the hunter might feel himself in the presence of a powerful, primal deity—one of the great animal spirits that was among humanity's first gods.

This is a highly speculative scenario, of course, but in neurological terms it is a compellingly plausible one. It is also consistent with the claims made by all the world's religions, that their spiritual origins can be traced to an epic mystical encounter with some incarnation of fundamental truth.

The Catholic monk and mystic Wayne Teasdale says in his book *The Mystic Heart*,

> Each great religion has a similar origin: the spiritual awakening of its founders to God, the divine, the absolute, the spirit, Tao, boundless awareness. We find it in the experience of the rishis in India; the Buddha in his experience of enlightenment; in Moses, the patriarchs, the prophets, and other holy souls of the biblical tradition. It is no less present in Jesus' inner realization of his relationship with his father. . . . And it is clear in the Prophet Mohammed's revelation experience of Allah through the mediation of the Archangel Gabriel.[7]

All the great scriptures make the same point: Fundamental truth has been revealed to human beings through a mystical encounter with a higher spiritual reality; mysticism, in other words, is the source of the essential wisdom and truth upon which all religions are founded. But before religions can begin, mystical experiences must be interpreted in rational terms, and the ineffable insights they bestow must be translated into specific beliefs.

For example, if the hunter's reaction to his spiritual encounter were similar to the reactions of latter day mystics (as neurology suggests it would be) it would overpower him with hope, reassurance, and indescribable joy, and he would no doubt race back to his clan to share the transforming revelation that great and beneficial powers exist in the world.

There's no way to know what the hunter's hungry clan mates would make of his strange story, but it wouldn't be surprising if, like mystics through the ages, he was greeted with skeptical suspicion. Let's imagine, though, that a few days later, hunters from the clan stumbled upon a small herd of deer and made their first kill in weeks. Our hunter would insist it was a gift from the Great Stag, intended as a display of his goodness and power. Others might be inclined to agree; in such a case, the story of the stag would acquire the dimensions of myth, and the clan mates would be compelled to elaborate upon it.

They might try to fathom the essential nature of this mysterious being, for example, to understand what it might want from humans, and what its spiritual powers might be. They might discuss the attributes of its character: Is it just? Wrathful? Reliable? Forgiving? Perhaps they would invent stories to explain the particulars of its existence: Where it dwells, for example, and how it came to be.

They would, in other words, establish a primitive theology, which sooner or later would address the questions upon which even the most sophisticated theologies are based: How can we please this god? How can we persuade him to protect us from the terrors of existence? How can we, through the powers of this spirit, seize the opportunity to control the most crucial aspects of our lives?

They might try to win the spirit's favor by making sacrificial offerings of animals or food. Anthropological evidence, in fact, suggests that religions, in their crudest forms, began as the attempt to influence powerful spirits with sacrificial gifts. The act of sacrifice is based upon the assumption of a contractual agreement between humans and the higher powers they believe in. This assumption is a defining characteristic of religion, and the component that lifts religion above the primitive level of magic. In magic, for example, spells and chants are intended to directly influence nature and other humans—to bring rain, heal the sick, curse an enemy and so on, without help from any spiritual intermediary.

In sophisticated religions, believers fulfill their end of the contract through worship, faith, obedience, and prayer. In return, they receive protection from evil, the absolution of sin, deliverance from worldly suffering, and some form of union with the divine. Our prehistoric hunters may have struck a simpler bargain, hoping only that their sacrifices would appease the spirit of the stag, who would reward them with abundant success in the hunt. But even this simple arrangement would have granted them a powerful sense of control over the existential uncertainties of their lives. The optimism and empowerment granted by this sense of control surely would have improved their psychological outlook, and given them a definite edge in the evolutionary struggle to survive.

In time, they would search for more sophisticated ways to interact with the spirit world. Perhaps intuition would lead them to dance in deerskins, hoping to coax the spirit of the stag into their midst. The religious practices of many hunting cultures have been based on such rituals, and whatever the spiritual intention of such ceremonies might be, they would almost certainly yield some positive, pragmatic results.

For example, the clan might begin to define itself by the rituals they perform, and the spirit who is the focus of those ceremonies. They might begin to think of themselves as the Clan of the Great

Stag and to draw upon their common beliefs as a source of identity and social cohesion. Ritual behaviors would also have a more direct effect upon survival rates, by stimulating the body's autonomic system. Slow rituals, for example, would stimulate a quiescent response, which could lead to various health benefits. Vigorous rituals would produce the beneficial advantages of exercise and even hone their crucial hunting skills.

Their religion would serve to strengthen bonds between individuals and to encourage more peaceful and productive interaction in the community at large. Stronger social groups, of course, would mean better lives for clan members, which might ultimately result in higher rates of survival as well.

Clearly, the physical, psychological, and social advantages of religious belief and behaviors would give religiously inclined humans a significant edge in the struggle for evolutionary survival. All these benefits, we believe, are a result of religion's power to grant humans a sense of control, which, in turn, is rooted in the mind's capacity for transcendent experience. When the hunter mystically encountered the spirit of the Great Stag, for example, he knew in his bones he was in the presence of something awesome and real. The neurological machinery, which made the experience possible, would allow him no other interpretation.

That same neurological machinery would be set in motion to a milder degree by the rhythmic behaviors of ritual. By participating in ritualistic ceremonies, the hunter's clan mates would personally experience a range of unitary states. Some of these states could be profound, but even at milder stages they would provide the clan mates with a taste of the spiritual union experienced by the hunter, lending credence to the hunter's story, and giving them firsthand reason to believe there was a spiritual presence in their midst.

In simpler terms, the credibility of religion, and especially its ability to grant a sense of control, is based in mystical experience.

As we've seen, neurology makes it clear that spiritual insights are born in startling moments of mystical transcendence. As we struggle to make sense of these perceptions, to identify their attributes and intentions, and to understand what our relationship with them should be, we develop the beliefs, traditions, and behaviors that are collectively known as religions.

Each religion may find its own definition of truth and chart a different course toward a union with the divine. A hundred human variables—history, geography, ethnicity, even politics—may help shape its final form. In every case, however, the authority of that religion and the essential realness of its God are rooted in transcendent experiences of mystical union, whether mild or extremely powerful.

If religions truly do arise from mystical insights, and if religious behaviors are as healthy as the research seems to show, then it's very likely that natural selection would favor a brain equipped with the neurological machinery that makes religious behavior more likely. There is no reason to suggest that this machinery evolved specifically for the purpose of mystical experience. As we explained earlier in the book, we believe the neurology of transcendence borrows the neural circuitry of sexual response—but the strong survival advantages of religious belief make it very likely that evolution would enhance the neurological wiring that makes transcendence possible. This inherited ability to experience spiritual union is the real source of religion's staying power. It anchors religious belief in something deeper and more potent than intellect and reason; it makes God a reality that can't be undone by ideas, and that never grows obsolete.

This does not mean, of course, that spiritual insights are always interpreted in healthy ways: The disastrous end of certain doomsday cults, for example, and the psychological damage wrought by dogmas of guilt and fear, give ample evidence that it is not.

It also does not mean that it is neurologically "abnormal" to be indifferent to religion, or that the claims of any specific religion are necessarily true. It only means that humans have a genetically inherited talent for entering unitary states, and that many of us interpret these states as the presence of a higher spiritual power.

A WINDOW TO GOD?

There's little doubt that the transcendent states from which religions rise are neurologically real—brain science predicts their occurrence, and our imaging studies, as well as others, have actually captured them on film. The deeper question is: Are these unitary experiences merely the result of neurological function—which would reduce mystical experience to a flurry of neural blips and flashes—or are they genuine experiences which the brain is able to perceive? Could it be that the brain has evolved the ability to transcend material existence, and experience a higher plane of being that actually exists?

The mystics certainly insist that they have experienced just such a reality; a realm of being more real than the material world that we trust without question, a dimension with no sensation of space, no passage of time, no clear boundaries between the self and the universe, and ample room for the actual possibility of God.

Science and common sense, on the other hand, tell us such a thing is not possible: Nothing can be more real than the material universe within which all real things are contained. Our own scientific inquiry, in fact, began with this assumption. But science has surprised us, and our research has left us no choice but to conclude that the mystics may be on to something, that the mind's machinery of transcendence may in fact be a window through which we can glimpse the ultimate realness of something that is truly divine. This conclusion is based on deductive reason, not on religious faith—it is a terrifically unscientific idea that is ironically consistent with careful, conventional science—but before it will make any sense, we must second-guess all our assumptions about material reality, and understand how the mind decides what is essentially and fundamentally real.

8

REALER THAN REAL

The Mind in Search of Absolutes

There is a theory which states that if ever anybody discovers exactly what the Universe is for and why it is here, it will instantly disappear and be replaced by something even more bizarre and inexplicable. There is another theory which states that this has already happened.

—Douglas Adams
The Restaurant at the End of the Universe

The universe certainly is strange, but to the average rationally minded individual, nothing is stranger, no concept can be more bizarre and inexplicable, than the claims of the mystics that another plane of being exists that is literally more real than the reality of the material universe. The transcendent reality they describe, in fact, would absorb the material world, along with the subjective self that perceives it, into the spiritual All, or the mystical Nothing, depending upon your metaphysical point of view.

Common sense, which tells us that nothing can be more real than the ground we walk on, or the chair in which we sit, compels us to reject this mystical reality as nonsense. A fair examination of mystical experience, however, shows that this is not so easy to do. As we've seen, mystics are not necessarily the victims of delusion, rather, their experiences are based in observable functions of the brain. The neurological roots of these experiences would render them as convincingly real as any other of the brain's perceptions. In this sense, the mystics are not talking nonsense; they are reporting genuine, neurobiological events.

This is the conclusion to which our research draws us; it forces us to ask a provocative question about the ultimate nature of human spirituality: Can all spirituality and any experience of the reality of God be reduced to a fleeting rush of electrochemical blips and flashes, racing along the neural pathways of the brain? Based upon our current understanding of the manner in which the brain turns neural input into the perceptions of human experience, the simplest answer is *yes*.

Are we saying, then, that God is just an idea, with no more absolute substance than a fantasy or a dream? Based upon our best understanding of how the mind interprets the perceptions of the brain, the simplest answer is *no*.

Our own brain science can neither prove nor disprove the existence of God, at least not with simple answers. The neurobiological aspects of spiritual experience support the *sense* of the realness of God. Yet we interpret and funnel that which our brain tells us is real through our subjective self-awareness. So, before we look further at the functions of the brain that links us to God, we need to discuss what the brain does to tell us that something is real—and why we believe it.

THE SCIENCE OF THE MYSTICS

We can most likely agree that there are two kinds of reality: the solid, objective external reality that we think of as "the world," and the inner, subjective sense of reality that we attribute to "the self." Based on everyday experience, we can't really dispute the realness of either one. Nor can we dispute the fact that one is essentially different from the other. If we agree that they are fundamentally different, however, and if they are the only two ways in which reality can exist, then logically only one of them can represent reality in its more fundamental form. In other words, either the objective external world or our subjective awareness of that world and the sense of self must be the real reality—the primary, ultimate reality. By definition, ultimate reality must be the source of everything that is real, so subjective and objective reality cannot both be true. One must be the source of the other.

Philosophers have struggled for centuries to understand the relationship between subjective and objective reality. And they are struggling still. Yet most of us are able to coexist comfortably in both. It is beyond the scope of this book, and far beyond the aims of this chapter, to chronicle the milestones in this epic intellectual quest, or to summarize their endless and often impenetrable arguments. Our goal here is to understand, as accurately and logically as possible, what the neurobiological "realness" of spiritual experience might imply. Since we hope to reach this understanding in a way that will satisfy the empirical demands of science, we'll begin by discussing the concept of reality that has provided the foundation for centuries of scientific thought.

WILL THE REAL REALITY, PLEASE . . .

In simplest terms, "scientific," or objective, reality is based on the belief that nothing is more real than the material world. In this view, external reality—the physical, material universe—is primary fundamental reality. Everything that is real, exists in, or has risen from, the material elements and forces of the universe. Even the human brain, and the subjective mind it makes possible, are material in nature, having evolved, like all biological systems, from some primordial ooze.

Mystics, however, have different ideas about what is fundamentally real. They believe they have experienced a primary reality that runs deeper than material existence—a state of pure being that encompasses the lesser realities of the external world and the subjective self. Science rejects this claim, not only because it holds that nothing in existence is more real than the reality of matter, but also because it cannot accept that something other than science, especially something as subjective and unmeasureable as mystical experience, can yield useful truth about what is fundamentally real.

The organizing principle of science declares that everything that is real can be measured, and scientific methods are the only measurements that count. So whatever can't be measured, weighed, counted, scanned, or otherwise analytically understood by scientific methods cannot, with any confidence, be called real. Science alone can recognize reality; Sigmund Freud said as much when he wrote, "Science is not an illusion! An illusion it would be to suppose that what science cannot give we can get elsewhere."[1]

In other words, the reality of the mystics cannot be considered real because it cannot be verified scientifically. We authors, as scientists, might have readily embraced this same conclusion had we not been convinced, by our own research, that the claims of the mystics just might be true. Science and mysticism, of course, are strange bedfellows, so let's retrace the steps that led us to this iconoclastic conclusion.

Gene and I began, as all scientists do, with the fundamental assumption that all that is really real is material. We regarded the brain as a biological machine, composed of matter and created by evolution to perceive and interact with the physical world.

After years of research, however, our understanding of various key brain structures and the way information is channeled along neural pathways led us to hypothesize that the brain possesses a neurological mechanism for self-transcendence. When taken to its extreme, this mechanism, we believed, would erase the mind's sense of self and undo any conscious awareness of an external world.

This hypothesis was later supported by our SPECT scan studies, which began to shed light on the neurological correlates of spiritual experience. In the narrowest scientific view, it would be possible to believe that we had reduced all spiritual transcendence—from the mildest case of religious uplift, to the profound states of union described by mystics—to a neurochemical commotion in the brain.

But our understanding of the brain would not allow us to rest with that conclusion. We knew, after all, that everything the mind experiences is tracked in the brain. A SPECT scan of an opera lover listening to Puccini, for example, would reduce "Nessun Dorma" to multicolored blotches, but that would not diminish the beauty of the aria. The music, and the enjoyment it provided, would still be very real. The memory of the music, too, and the emotional pull of the tragedy of *Turandot*, are real. Even if you were to "play" the music and drama again only in your mind, many of the same parts of the brain would be reactivated. Perhaps even your body would get the same goose bumps evoked by Puccini's heartbreaking lyrical melody, its crescendos and pianissimos. You would clearly be hearing the music, but only inside your head. Yet the existence of the music and its nonverbal power are still, neurologically, quite real.

All perceptions exist in the mind. The earth beneath your feet, the chair you're sitting in, the book you hold in your hands may all seem unquestionably solid and real, but they are known to you only as secondhand neurological perceptions, as blips and flashes racing along the neural pathways inside your skull. If you were to dismiss spiritual experience as "mere" neurological activities, you would also have to distrust all of your own brain's perceptions of the material world. On the other hand, if we do trust our perceptions of the physical world, we have no rational reason to declare that spiritual experience is a fiction that is "only" in the mind.

At this point in our research, science had brought us as far as it could, and we were left with two mutually exclusive possibilities: either spiritual experience is nothing more than a neurological construct created by and contained within the brain, or the state of absolute union that the mystics describe does in fact exist and the mind has developed the capability to perceive it.

Science offers no clear way to resolve this question. But we knew that at least we had found a new framework for understanding the phenomenon of mystical experience, no matter what reality these brain states ultimately represent.

The transcendent state we call Absolute Unitary Being refers to states known by various names in different cultures—the Tao, Nirvana, the *Unio Mystica*, Brahman-atman—but which every persuasion describes in strikingly similar terms. It is a state of pure awareness, a clear and vivid consciousness of *no-thing*. Yet it is also a sudden, vivid consciousness of *everything* as an undifferentiated whole.

Although mystics report that this state of ultimate being cannot be understood through reason, or even rationally described, that hasn't stopped legions from trying. Most written accounts, while intriguing, are hopelessly perplexing and often contradictory. This does not mean that they are not true, however, or that they do not describe reality accurately.

Consider, for example, the words of the modern Zen master Huang Po, describing the ultimate state of being he calls One Mind:

> All the Buddhas and all sentient beings are nothing but One Mind, beside which nothing exists. This Mind, which is without beginning, is unborn and indestructible. It is not green or yellow, and has neither form nor appearance, it

> does not belong to the categories of things which exist or do not exist, nor can it be thought of in terms of new or old. It is neither long nor short, big nor small, for it transcends all limits, measures, names, traces, and comparisons. Only awake to the One Mind.[2]

It is difficult for the rational mind to accept these cryptic pronouncements as fact: The One Mind is uncreated; it is *not* nonexistent, but at the same time cannot be said to exist; it transcends all limits and comparisons; and, outside of this One Mind, nothing else is real.

To anyone who has not experienced such a unitary state mentally and physically, the meaning of these concepts may be difficult to grasp. Ironically, the state must be felt to be believed by both body and mind, even though both body and mind are transcended by it. But mystics insist that it is completely possible to understand—and attain—if we set aside our subjective disbelief. Absolute Unitary Being is described as a state without time, space, and physical sensations; with no discrete awareness of any material reality at all. Ironically, again, the attainment of absolute unitary being requires a mental journey into the deepest parts of the self, yet those who have reached this ultimate state agree that subjective self-awareness utterly vanishes once Absolute Unitary Being has been achieved. So, to get to this state we have to use the mind to get beyond the mind. The mind has to get out of its own way.

This obliteration of the self may be the most difficult concept for the rational mind to comprehend. We are so at home in the state of subjective awareness that we find it hard to fathom how a mind that does not contain a specific self can be much of a mind at all. It is possible, however, for awareness to exist without the subjective focus of a self. In fact, we all begin life with selfless minds.

All human babies are born with the neurological potential to form a self, but they do this through living and experiencing the world as they grow older. The development of the self also requires the development of certain neural connections in the brain.

HOW THE MIND MAKES THE SELF

The process by which the self is derived is a mystery, but we believe it might arise through a process of reification—the ability to convert a concept into a concrete thing, or, more succinctly, to bestow upon something the quality of being real or true. In its neurological definition, the term refers to the power of the mind to grant meaning and substance to its own perceptions, thoughts, and beliefs, and to regard them as meaningful.

For example, it's the reifying power of the mind, granted primarily through the function that we call the abstractive operator, within the inferior parietal lobe, that allows you to look at a watermelon, a berry, a rock-hard coconut, or a soft fuzzy peach and recognize them all as "fruit." But reification not only allows you to recognize the essential sameness of things, it also enables you to perceive them as substantial and real. We previously described this brain function as the existential operator.

Making the world concrete—the process of reification—may play an important role in the neural development of the self. The young brain starts dealing with the world beyond its internal feelings and operations by making sounds and physical actions. At first, the brain puts out these behaviors, then it observes and analyzes and inputs them as new information. Eventually, the brain identifies, or *reifies*, these actions, thoughts, and feelings as the self.

For example, imagine a baby laughing in his crib. The moment the laughter is audible it becomes a part of the external world and is delivered to the baby's brain as new sensory input, which the brain recognizes as the result of its own neurological function. At the same time, the baby's brain perceives the presence of the mother, who responds to the sound of laughter by clapping her hands in delight. The child's brain can find no correlation within itself for this behavior; in other words, it recognizes the mother's behavior as something other than itself.

By comparing these perceptions, the baby's brain begins to recognize two general categories of sensory input—the first is input resulting from his behavior, the second is input from behaviors he does not generate or control. We believe that the perception of these categories is the first step in the brain's inclination to draw a line between the inner reality of the self and the external reality of the world.

As his experience with the outside world continues, the child's brain is able to recognize more and more behaviors that seem to be his own. Eventually, those various independent functions—thoughts, emotions, intentions, actions, and memories—are all categorized as a single, distinct, meaningful construct. In other words, they become reified into the specific, familiar, enduring, and highly personal "self."

This process of how the brain makes the self is theoretical, of course, but highly probable. And it makes some important points: The self is not the same as the mind. The mind exists before the self and, in one way or another, it supplies the essential memories, emotions, and other component parts from which the self is assembled.

If these components could somehow be undone, the self would come unraveled. We believe that this is exactly what happens when the orientation association area, as well as other areas that might help to provide a sense of the self, becomes deafferented—deprived of new sensory input. These areas are also cut off from the memories, emotions, and patterns of behavior that the mind recognizes as the self. Deafferentation does not deprive the mind of awareness, it simply frees that awareness of the usual subjective sense of self, and from all sense of the spatial world in which that self could be.

The result of such a lack of input, almost certainly, would be a state of pure awareness, an awareness stripped of ego, focused on nothing, oblivious to the passage of time and physical sensation. This awareness would be neurobiologically incapable of differentiating between subject and object, between the limited personal self and the external, material world. It would perceive and interpret reality as a formless unified whole, with no limits, no substance, no beginning, and no end.

All the assembled constructs of the conscious mind—the emotions, memories, thoughts, and unformed intuitions by which we know our selves—would come undone, and dissolve into this underlying pure awareness, which would be our deepest, truest self, the universal self the mystics describe.

"The way in which you come to know you are you," says contemporary Buddhist minister Leslie Kuwamara, "is by a process of elimination." For Taoist sage Li Po, the evaporation of the illusory self provides the clarity we need if we are to recognize, without question, what is truly and, most simply, real:

> The birds have vanished into the sky,
> and now the last cloud drains away.
>
> We sit together, the mountain and me,
> until only the mountain remains.[3]

While our neurological model offers a plausible explanation of how we experience the mystical state of pure awareness, it proves nothing about the ultimate nature of Absolute Unitary Being. It does not explain whether absolute being is nothing more than a brain state or, as mystics claim, the essence of what is most fundamentally real. Yet our work has convinced us that the mystics, at the very least, are not delusional or psychotic: They are certain beyond a shadow of a doubt that their experiences are real.

Since no empirical method can objectively test that realness, we have to turn instead to the more subjective approach of the philosophers. After centuries of inquiry, philosophers have come to suggest that true reality possesses an unmistakable quality. The Stoics defined this quality as the *phantasia catalyptica*; certain modern German thinkers call it *Anweisenheit*, and phenomenologists describe it as intentionality.

All these phrases mean that what's real simply feels more real than what's not. This may seem an unsatisfyingly soft standard, but it is the best guidance that the greatest minds and experts have produced.[4] In most cases, it works quite well, and all other approaches to this problem are ultimately reduced to this assertion.

For example, dreams can feel remarkably real while the dream state persists, but when we wake, the insubstantial nature of the dream state becomes immediately clear. We consider waking reality a higher degree of reality than the reality of dreams, because it feels more convincingly real. We could say the same about the reality of daydreaming, or of various hallucinatory states. Each of these realities may seem quite real while it persists, but when it ends, and we consider it contrasted with ordinary, or "baseline" reality, we dismiss it as something less than real.

The realness of the material world, therefore, is made clear to us when we compare it with other states. Since most of us have never experienced a state more real than the one our mind portrays for us every day, we have no reason to suspect that any higher reality exists beyond our subjective awareness of the material world. More important, we have no experimental reason to believe that any higher reality is even possible.

Those who have experienced advanced states of mystical unity, however, claim that these states do feel like a higher reality. Passionately and consistently, with a preponderance of agreement that stretches across history and embraces all faiths, they insist that when compared to our baseline sense of reality, Absolute Unitary Being is more vividly, more convincingly real.

The mystics' claims are supported by some of the greatest scientists of the century—rational thinkers who have peered deeper than most into the workings of the universe and the mind and have described states of transcendent spiritual awareness in words that mirror the accounts of the gurus, shamans, and saints in remarkably specific detail. Robert Oppenheimer, Neils Bohr, Carl Jung, and John Lilly are among the prominent scientific figures whose work has revealed to them a unity and purpose in the workings of the universe that transcend the material world.

Most impressive, perhaps, are the accounts of physicists Albert Einstein and Edwin Schrödinger, two great thinkers who have perhaps understood the nature of scientific reality most clearly. The theories for which these men are best known—Einstein's relativity and Schrödinger's quantum mechanics—provide our most basic understandings of how the universe functions; much of our understanding of physical reality is based upon them. Scientifically, Einstein and Schrödinger did not agree on the fundamental nature of existence—Einstein could never accept the pretzel logic of quantum theory—but their lifelong preoccupation with the gears and cogs and forces that gave birth to, and sustain the world, led each of them to a deeper understanding of the essence of things. On that profound level, they seem to find agreement.

For Einstein, this understanding expressed itself as a longing for something larger than himself, an experience he referred to a "cosmic religious feeling":

> It is very difficult to explain this feeling to anyone who is entirely without it, especially as there is no anthropomorphic conception of God corresponding to it. The individual feels the nothingness of human desires and aims and the

> sublimity and marvelous order which reveal themselves both in Nature and in the world of thought. He looks upon individual existence as a sort of prison and wants to experience the universe as a single significant whole.[5]

For Schrödinger, the wholeness Einstein longs for is satisfied in the apprehension of the oneness of all things:

> Inconceivable as it seems to ordinary reason, you—and all other conscious beings as such—are all in all. Hence, this life of yours you are living is not merely a piece of the entire existence, but is in a certain sense the whole. . . . Thus, you can throw yourself flat on the ground, stretched out upon Mother Earth with a certain conviction you are one with her and she with you. You are as firmly established, as invulnerable as she, indeed a thousand times firmer and more invulnerable.[6]

In the opinion of biologist Edwin Chargaff, all real scientists are driven by the mysterious intuition that something immense and unknowable dwells in the material world: "If [a scientist] has not experienced, at least a few times in his life, this cold shudder down his spine, this confrontation with an immense, invisible face whose breath moves him to tears, he is not a scientist."[7]

It seems that even the self-avowed agnostic Carl Sagan was not immune to the "mysterious intuition" Chargaff describes. In his novel *Contact*, his main character, scientist Ellie Arroway, describes a profound personal experience in terms every ancient mystic would recognize:

> I had an experience I can't prove. I can't even explain it, but everything I know as a human being, everything that I am tells me that it was real. I was part of something wonderful,

> something that changed me forever; a vision of the Universe that tells us undeniably how tiny, and insignificant, and how rare and precious we all are. A vision that tells us we belong to something that is greater than ourselves. That we are not, that none of us is, alone.[8]

Logic suggests that what is less real must be contained by what is more real, just as a dream is contained within the mind of a dreamer. So, if Absolute Unitary Being truly is more real than subjective or objective reality—more real, that is, than the external world and the subjective awareness of the self—then the self and the world must be contained within, and perhaps created by, the reality of Absolute Unitary Being.

Again, we cannot objectively prove the actual existence of Absolute Unitary Being, but our understanding of the brain and the way it judges for us what is real argues compellingly that the existence of an absolute higher reality or power is at least as rationally possible as is the existence of a purely material world.[9]

Although the notion of a reality more real than the one in which we live is difficult to accept without personal experience, when the mind drops its subjective preoccupation with the needs of the self and the material distractions of the world, it can perceive this greater reality. Mystical reality holds, and the neurology does not contradict it, that beneath the mind's perception of thoughts, memories, emotions, and objects, beneath the subjective awareness we think of as the self, there is a deeper self, a state of pure awareness that sees beyond the limits of subject and object, and rests in a universe where all things are one.

> Siddhartha listened. He was now listening intently, completely absorbed, quite empty, taking in everything. He felt that he had now completely learned the art of listening. He

> had often heard all this before, all these numerous voices in the river, but today they sounded different. He could no longer distinguish the different voices—the merry voice from the weeping voice, the childish voice from the manly voice. They all belonged to each other: the lament of those who yearn, the laughter of the wise, the cry of indignation and the groan of the dying. They were all interwoven and interlocked, entwined in a thousand ways. And all the voices, all the goals, all the pleasures, all the good and evil, all of them together was the world. All of them together was the stream of events, the music of life. When Siddhartha listened attentively to this river, to the song of a thousand voices, when he did not listen to the sorrow or the laughter, when he did not bind his soul to any one particular voice and absorb it in his Self, but heard them all, the whole, the unity, then the great song of a thousand voices consisted of one word.[10]

The wisdom of the mystics, it seems, has predicted for centuries what neurology now shows to be true: In Absolute Unitary Being, self blends into other; mind and matter are one and the same.

9

WHY GOD WON'T GO AWAY

The Metaphor of God and the Mythology of Science

The one whom I bow to only knows to whom I bow
When I attempt the ineffable Name, murmuring Thou
And dream of Phaedian fancies and embrace in heart
Symbols (I know) which cannot be the thing thou art.
Thus always, taken at their word, all prayers blaspheme
Worshipping with frail images of folk-lore dream,
And all in their praying, self-deceived, address
The coinage of their own unquiet thoughts, unless
Thou in magnetic mercy to Thyself divert
Our arrows, aimed, unskillfully, beyond desert;
And all are idolators, crying unheard
To a deaf idol, if thou take them at thy word.
Take not, O Lord, our literal sense. Lord, in thy great,
Unspoken speech our limping metaphor translate.

—C. S. Lewis
"A Footnote to All Prayers"

The prominent Christian apologist C. S. Lewis was never noted for his mystical sensibilities—he was an Oxford don who relied upon intellect and meticulous scholarship to argue in defense of his faith, and whose writings have been embraced by millions of traditional, literalistic Christians around the world. In the poem above, however, he steps beyond mainstream orthodoxy to echo the essential transcendent wisdom of the mystics, that God is beyond all comprehension and description, and that all literal interpretations of his unknowable nature can never be more than symbols pointing toward a deeper, more mysterious truth.

The *unknowableness* of God is a defining principle for the mystically inclined religions of the East. Buddhism and Taoism, for example, leave little room for any personified deity. Even Hindus, who worship specific personalized deities, understand that these specific, identifiable gods are representations of the one supreme Godhead, Brahman, who exists beyond form and description, and for whom "all illustrations are inadequate and truth is beyond words."[1]

The concept of an unknowable, ungraspable God is more difficult for the monotheistic religions of the West. Judaism, Christianity, and Islam are all founded upon the revelation of a God who is a distinct Supreme Being: a specific supernatural entity, set apart from the natural world, with a name, a history, and a specific agenda for his people. The Western God speaks to the world through scriptures and prophets, he has moods and emotions, he is believed to be dramatically and empirically *real.* Such a strong divine personality does not lend itself easily to transcendent interpretations, but the mystics of the three great Western faiths have consistently and collectively insisted, in agreement with the teachings of the East, that the ultimate essence of God is far beyond the reach of human comprehension.

The high God of Kabbalistic mysticism, for example, sounds very much like the Hindu Brahman: a divine concept beyond the reach of human understanding, with no form or limits, and no distinct personal attributes whatsoever. The Kabbalah calls this God *Ein-Sof*, which translates literally as "endless" or "infinite."

In *The Essential Kabbalah*, Judaic scholar Daniel Matt explains that God as *Ein-Sof* is beyond limits and comparisons. "Anything visible, and anything that can be grasped by thought is bounded," he says. "Anything bounded is finite. Anything finite is not undifferentiated. Conversely, the boundless is called *Ein-Sof*, Infinite. It is absolute undifferentiation in perfect, changeless oneness. . . ."

Islamic mystics have also understood the boundless ineffability of God, and have expressed the futility of trying to explain his unknowable nature. According to Rabi'a al-Adawiyya an eighth-century Islamic saint:

> The one who explains, lies.
> How can you describe the true form of Something
> In whose presence you are blotted out?
> And in whose being you still exist?[2]

The spiritual experiences of Christian mystics has likewise led them to conclude that the urge to understand God as a specific, literal being only leads us astray. "If you wish to be perfect and without sin," says the Catholic mystic Meister Eckhart, "then do not prattle about God. Also, you should not wish to understand anything about God, for God is beyond all understanding. A master says: 'If I had a God that I could understand, I would not regard him as God.' If you understand anything about him, then he is not in it, and by understanding something of him, you fall into ignorance. . . ."[3]

The conclusions of the mystics seem clear: God is by his nature unknowable. He is not an objective fact, or an actual being; he is, in fact, *being* itself, the absolute, undifferentiated oneness that is the ground of all existence. When we understand this truth, the mystics claim, all religions connect us to this deeper, divine power. If we fail to understand it and we cling to the comforting images of a personal, knowable God—a God who exists entirely apart from the rest of creation as a distinct, individual being—we diminish the ultimate realness of God, and reduce his divinity to the stature of the small, "deaf idol" that Lewis's poetry so poignantly describes.

The mystics claim that the true nature of God can only be known through a direct, mystical encounter. Evelyn Underhill explains that "Mysticism, in its pure form, is the science of ultimates, the science of union with the Absolute and nothing else," and that "the mystic is the person who attains this union, not the person who talks about it. Not to *know about*, but to *Be*, is the mark of the real initiate."[4]

That doesn't mean, however, that the rest of us cannot appreciate the insights the mystics have shared. "Even if we are incapable of the higher states of consciousness achieved by a mystic," says Karen Armstrong, "we can learn that God does not exist in any simplistic sense, for example, or that the very word 'God' is only a symbol of a reality that ineffably transcends it."

For centuries, the existence of such a reality has been supported only by the claims of mysticism. Science has traditionally rejected these claims, but our work suggests that in the form of Absolute Unitary Being, the spiritual union the mystics describe feels at least as solidly and as literally real as any other experience of reality. The neurological and philosophical correlates of this conviction make it clear that Absolute Unitary Being would be a state of ultimate union and total undifferentiated oneness, a plane of existence in which all degrees of difference dissolve and comparisons become impossible. In Absolute Unitary Being, nothing is experienced but the pure and complete unity of all things, or of *nothings*. One thing cannot stand apart from another, so individual beings and objects cannot be perceived. The egotistical self cannot exist, because it has no *non-self* against which to define itself. In the same fashion God cannot be set apart from this ultimate oneness as an identifiable, personalized being—to do so would be to conceive of a God who is less than absolutely real.

The perception of an absolute reality therefore, would demand that God be more than a knowable being, and make it clear that all personifications of God are symbolic attempts to grasp the ungraspable. This does not mean, however, that personalized conceptions of God are meaningless, or necessarily untrue. Instead, the state of Absolute Unitary Being impresses upon anyone who experiences it, the realization that the God we can know is only a glimmer of a higher spiritual reality, in the same way, perhaps, that a single beam of light implies the glory of the sun. In this sense, all personalized incarnations of God are rooted in the perception of a larger reality, the deepest, most sublime sense of realness the mind is able to experience. In this ultimate realness, which lies beyond material existence and subjective experience, all conflicts are resolved and the fundamental promise of all religions is fulfilled—suffering ends, unity and bliss are eternal.

In *The Mystic Heart*, Wayne Teasdale addresses the apparently irreconcilable conflict between belief in a personal God, and the acknowledgment of an impersonal higher reality. "In resolving these seemingly conflicting views," he says,

> I think we are going to discover in coming years a more adequate view of the divine—something that can be verified in mystical experience—includes both personal and *trans*personal reality. God is both a loving presence, compassionate, wise, kind, and merciful, and an impersonal principle of ultimate condition of consciousness, the basis of karma, *shuyata* or emptiness, and *nirvana*. They represent two sides of the same source, two fundamental insights, two mystical realizations of the ultimate mystery.[5]

In other words, various incarnations of God are metaphorical interpretations of the same spiritual reality—the reality experienced as Absolute Unitary Being. When we realize that any specific conception of God is a piece of this larger puzzle, rooted in a mystical understanding of what's fundamentally real, then all religions become siblings, all faiths become true, and all incarnations of God can be understood as real.

But if we reject this crucial insight, if we cling to the image of God as a literal, knowable being, distinct and independent of the rest of reality, then we are left, at best, with a God that Lewis's poem describes as a "limping metaphor," as a symbol "which cannot be the thing thou art." At worst, we create a God who leads us away from unity and compassion, toward division and strife.

In terms of its impact upon human history, the concept of a personalized God is not necessarily negative. In *A History of God*, Karen Armstrong points out that the presence of just such a God, in fact, has made profoundly positive contributions to Western culture. "The personal God has helped monotheists to value the sacred and inalienable rights of the individual and to develop an appreciation of human personality," she says. "The Judeo-Christian tradition has thus helped the West to acquire the liberal humanism it values so highly."

But the same personal God, Armstrong warns, can become a grave liability. "[A personalized God] can be a mere idol carved in our own image," she points out,

> a projection of our limited needs, fears, and desires. We can assume that he loves what we love and hates what we hate, endorsing our prejudices instead of compelling us to transcend them. When he seems to fail to prevent a catastrophe or seems even to desire a tragedy, he can seem callous and cruel. A facile belief that a disaster is the will of God can make us accept things that are fundamentally unacceptable. The very fact that, as a person, God has a gender is also

> limiting: It means that the sexuality of half the human race is sacralized at the expense of the female and can lead to a neurotic and inadequate imbalance in human sexual mores. A personal God can be dangerous, therefore. Instead of pulling us beyond our limitations, "he" can encourage us to remain complacently within them; "he" can make us as cruel, callous, self-satisfied and partial as "he" seems to be. Instead of inspiring the compassion that should characterize all advanced religions, "he" can encourage us to judge, condemn, and marginalize.[6]

The God Armstrong describes is the God of witch-hunts, inquisitions, holy wars, fundamentalist intolerances, and countless other forms of religious persecution, all carried out with the confident presumption of divine endorsement. The authority to commit such atrocities is rooted in the assumption, made by believers, that their God is the only God, and their religion is the single, exclusive path to truth. As God's chosen people, they have the right, even the obligation, to oppose the "enemies of God" which is how they would describe the less holy, less *human* individuals who do not share their literal beliefs.

History suggests that religious intolerance is primarily a cultural phenomenon, based in ignorance, fear, xenophobic prejudice, and ethnocentric chauvinism. We believe, however, that intolerance is rooted in something deeper than mere narrow-mindedness; we believe it is based in the same transcendent experiences that foster belief in the absolute supremacy of personalized, partisan gods.

Transcendent states, as we've seen, exist along a continuum of progressively higher levels of unitary being that ultimately leads to the point at which unity becomes absolute. In the state of absolute unity, there are no competing versions of the truth; there is only truth itself, so conflicting beliefs, or conflicts of any kind for that matter, are not even possible.

If however, a mystic falls short of absolute unity—if, in neurological terms, the deafferentation of the orientation area is not complete—then subjective awareness would survive, and the mystic would interpret the experience as an ineffable union between the self and some mystical other. We examined the neurobiology of just such a state—the *Unio Mystica*—in our discussion of active meditation.

Like all advanced unitary states, this mysterious union would have a profound sense of realness; the mystics would viscerally feel that he or she had stood in the presence of absolute reality. A Christian might call this truth Jesus, a Muslim might invoke the name al-Lah, in primal cultures it might be interpreted as some powerful spirit of nature, but in every case it is experienced as a spiritual truth that stands apart from and above all others.

We've seen that the "discovery" of such truth through mystical experience, provides believers with a powerful sense of control over the otherwise uncontrollable whims of fate. The presence of a powerful spiritual ally convinces believers that their lives are a part of some comprehensible plan, that goodness rules the world, and even that death can ultimately be conquered.

What makes these beliefs more than hollow dreams is the fact that the God that stands behind them has been verified, through a direct mystical encounter, as literal, absolute truth. Any challenge to the authenticity of that truth, therefore, is an attack not only upon ideas about God, but also upon the deeper, neurobiologically endorsed assurances that make God real. If God is not real, neither is our most powerful source of hope and redemption. There can be only one absolute truth; it is a matter of existential survival. All others are threats of the most fundamental kind, and they must be exposed as impostors.

In other words, the presumption of "exclusive" truth, upon which religious intolerance is based, may rise out of incomplete states of neurobiological transcendence. Ironically, when the process of transcendence is taken to the logical, and neurobiological, extreme, the mind is confronted with a state of absolute, uncompromising unity, in which all conflict, all contradictions, all competing variations of the Truth, disappear into harmonic, monolithic oneness. If we are right, if religions and the literal Gods they define are in fact interpretations of transcendent experience, then all interpretations of God are rooted, ultimately, in the same experience of transcendent unity. This holds true whether this ultimate reality actually exists, or is only a neurological perception generated by an unusual brain state. All religions, therefore, are kin. None of them can exclusively own the realist reality, but all of them, at their best, steer the heart and the mind in the right direction.

Human history shows that few religions have been willing to take such an inclusive and magnanimous view, and conflict has almost always been the norm. There are encouraging signs, however, that a number of important religious thinkers are beginning to accept the possibility, even the promise, that all religions share common spiritual bonds.

"Human beings naturally possess different interests and inclinations," says the Dali Lama.

> Therefore, it should come as no surprise that we have many different religious traditions with different ways of thinking and behaving. But this variety is a way for everyone to be happy. If we only have bread, people who eat rice are left out. With a great variety of food, we are able to satisfy everyone's different needs and tastes. And people eat rice because it grows best where they live, not because it is either any better or worse than bread.[7]

"Because all the world's religious traditions share the same essential purpose, we must maintain respect and harmony among them."[8] In his book *The Mystic Heart*, Catholic laymonk and scholar Wayne Teasdale describes what he sees as the beginnings of the very kind of interreligious respect and harmony the Dali Lama urges. Teasdale has coined the term "interspirituality" to describe this growing climate of dialogue and reconciliation between world religions, and to call attention to the essential spirituality that all religions share.

"Interspirituality is not about eliminating the world's rich diversity of religious expression," Teasdale says. "It is not about rejecting these traditions' individuality for a homogenous super-spirituality. It is not an attempt to create a new form of spiritual culture. Rather, it is an attempt to make available to everyone all the forms the spiritual journey assumes."

For Teasdale, that journey begins in the primal human urge for mystical union with some larger, all-embracing truth, a truth that humans sought long before religions existed to define it.

"For thousands of years before the dawn of the world religions as social organisms working their way through history, the mystical life thrived," Teasdale says. This mystical longing for connection with the divine, Teasdale believes, is at the heart of all genuine faith, and is the essence of all religions.

"The real religion of humankind can be said to be spirituality itself," claims Teasdale, "because mystical spirituality is the source of all the world's religions."

Teasdale's beliefs are based on faith, and on his personal experiences as a mystic, but his spiritual approach has led him to a conclusion that is also suggested by neurology: All religions arise from and are maintained by transcendent experiences, therefore, they all lead us, by different paths, toward the same goal of wholeness and unity, in which the specific claims of individual faiths converge into an absolute, undifferentiated whole.

For the proponents of interspirituality, the awareness of this fundamental unity has deep spiritual connotations, but it has profoundly pragmatic implications as well. As author and religion scholar Dr. Beatrice Bruteau argues in the preface to *The Mystic Heart*, mysticism may provide the world with its last, best hope for a happier future, by allowing us to overcome the greed, mistrust and self-protective fears that have led to so many centuries of suffering and strife.

"Consider that domination, greed, cruelty, violence, and all our other ills arise from a sense of insufficient and insecure *being*," says Bruteau.

> I need more power, more possessions, more respect and admiration. But it's never enough; the fear always remains. It comes from every side: from other people; from economic circumstances; from ideas, customs, and belief systems; from the natural environment; from our own bodies and minds. All these *others* intimidate us, threaten us, make us anxious. We can't control them. They are, to varying degrees, aliens. Our experience is: Where I am "I," they are "not-I."

According to Bruteau, mysticism allows us to transcend these egotistical fears. The awareness of mystical wholeness shows us that we are not so fundamentally alienated from one another and that, in fact, we do have all the *being* we need to be happy. When the appreciation of this mystical oneness rises to the surface, Bruteau says, "our motives, feelings, and actions turn from withdrawal, suspicion, rejection, hostility, and domination to openness, trust, inclusion, nurturance, and communion."

"This oneness—this freedom from alienation and insecurity—is the sure foundation for a better world," she says. "It means that we will try to help each other rather than hurt each other."

Anyone who reads the newspapers—anyone, for example, who has heard of Rwanda, Kosovo, Kashmir, or the Golan Heights—may find Bruteau's optimism poignantly naive. It's in our nature, it seems, to define survival as a matter of competition and conquest, as the survival of the fittest, as a ruthless game of dog-eat-dog. The human brain, after all, evolved primarily for the purpose of making us ferociously efficient competitors. We have a natural genius for defining threats, for naming enemies, and for fiercely protecting our own best interests, no matter how narrowly or selfishly we define them.

This does not mean, however, that we are condemned by human nature to live in a world of dissension and discord, because the same brain that inclines us toward egotistic excess also provides the machinery with which the ego can be transcended. In these transcendent states, whatever their ultimate spiritual nature may be, suspicion and dissension disappear into the peace and love of an indescribable unity. The transforming power of these unitary states, is what makes mysticism our most practical and effective hope for improving human behavior, believes Beatrice Bruteau. "If we could arrange energy from *within*," she says, "if we more often nurtured our companions and promoted their well-being, we would suffer much less. Rearranging energy from within is what mysticism does."

Generations may pass before human society is ready for such transforming ideas, but it is intriguing to know that if such a time should arrive, the brain will be ready, possessing the machinery it needs to make those ideas real.

The neurology of transcendence can, at the very least, provide a biological framework within which all religions can be reconciled. But if the unitary states that the brain makes possible are, in fact, glimpses of an actual higher reality, then religions are reflections not only of neurological unity, but of a deeper absolute reality.

In the same sense, neurology can reconcile the rift between science and religion, by showing them to be powerful but incomplete pathways to the same ultimate reality. The conflict between science and faith was propelled by the great discoveries of the scientific age, which arguably began when Galileo's observations verified Copernicus's view of the solar system. The revelation that the earth had not been lovingly set at the center of the universe by a doting, divine creator was a devastating blow to orthodox Christian doctrine at that time. To make matters worse, when the Church tried to silence Galileo by proclaiming him a heretic, it showed itself, in the eyes of many rational people, to be more concerned with dogma than with truth.

As centuries passed, and science and philosophy found more and more rational explanations for mysteries which once could only be explained by a divine presence, thinking people found it increasingly difficult to maintain their belief in God. Then in the mid-nineteenth century, science produced two revolutionary theories that seemed to make God irrelevant in the scientific age.

The first appeared in Charles Lyell's book *Principles of Geology*, published in 1830. Lyell's research showed that the contours of the natural landscape were shaped by geological forces, not by the hand of God, and that the earth was much older than Bible stories claimed. Twenty-nine years later, *The Origin of Species* was released, and the world was rocked by Darwin's revolutionary theories that life-forms evolved through impartial biological adaptation over a span of millions of years, and not in a single flash of divine creative energy.

In the midst of this scientific revelation, Nietzsche proclaimed God dead. It's important, however, to realize that the God he thought science had killed, the God that was no longer compatible with rational thinking, was the personal Creator God of the Bible. There is nothing that we have found in science or reason to refute the concept of a higher mystical reality.

This does not mean that mainstream science has opened its arms to the possibility of mystically-discovered reality. The authority of science, after all, is rooted in the assumption that material reality is reality in its highest form; that nothing is more real than the physical, material stuff of the universe. But even from a scientific perspective, the nature of material reality may be more slippery than common sense would suggest. Albert Einstein certainly thought so. In 1938, he expressed his belief that scientific interpretations of the physical world may not be as reliable as rational materialists would like to believe:

> Physical concepts are free creations of the human mind, and not, however it may seem, uniquely determined by the external world. In our endeavor to understand reality, we are somewhat like a man trying to understand the mechanism of a closed watch. He sees the face and the moving hands, even hears it ticking, but he has no way of opening the case. If he is ingenious he may form some picture of a mechanism which could be responsible for all the things he observes, but he may never be quite sure his picture is the only one which could explain his observations. He will never be able to compare his picture with the real mechanism and he cannot even imagine the possibility of the meaning of such a comparison.[9]

The best that science can give us is a metaphorical picture of what's real, and while that picture may make sense, it isn't necessarily true. In this case, science is a type of mythology, a collection of explanatory stories that resolve the mysteries of existence and help us cope with the challenges of life. This would be applicable even if material reality is, in fact, the highest level of reality, because despite science's preoccupation with objectively verified truth, the human mind is incapable of purely objective observations. All our perceptions are subjective by their nature, and just as there's no way to peek inside Einstein's watch, there's no way we can slip free of the brain's subjectivity to see what's really out there. All knowledge, then, is metaphorical; even our most basic sensory perceptions of the world around us can be thought of as an explanatory story created by the brain.

Science, therefore, is mythological, and like all mythological systems of belief, it is based on a foundational assumption: *All that is real can be verified by scientific measurement, therefore, what can't be verified by science isn't really real.*

This kind of assumption, that one system is exclusive arbiter of what is true, makes science and religion incompatible. If Absolute Unitary Being does, indeed, exist, then science and religion find themselves in a paradoxical situation: The more literally we take their own foundational assumptions, the deeper they are in conflict with each other, and the further they fall from ultimate reality. But if we understand the metaphorical nature of their insights, then their incompatibilities are reconciled, and each becomes more powerfully and transcendently real.

If Absolute Unitary Being is real, then God, in all the personified ways humans know him, can only be a metaphor. But as C. S. Lewis's poem suggests, metaphors are not meaningless, they do not point at nothing. What gives the metaphor of God its enduring meaning is the very fact that it is rooted in something that is experienced as unconditionally real.

The neurobiological roots of spiritual transcendence show that Absolute Unitary Being is a plausible, even probable possibility. Of all the surprises our theory has to offer—that myths are driven by biological compulsion, that rituals are intuitively shaped to trigger unitary states, that mystics are, after all, not necessarily crazy, and that all religions are branches of the same spiritual tree—the fact that this ultimate unitary state can be rationally supported intrigues us the most. The realness of Absolute Unitary Being is not conclusive proof that a higher God exists, but it makes a strong case that there is more to human existence than sheer material existence. Our minds are drawn by the intuition of this deeper reality, this utter sense of oneness, where suffering vanishes and all desires are at peace. As long as our brains are arranged the way they are, as long as our minds are capable of sensing this deeper reality, spirituality will continue to shape the human experience, and God, however we define that majestic, mysterious concept, will not go away.

NOTES

Chapter 1: A Photograph of God?

1. Because the "peak" moment in meditation is subjective, it is obviously very difficult to identify and even more difficult to measure. However, the peak state is also the most interesting because it is the state that carries the greatest spiritual meaning and has the most impact on the individual. Peak experiences may be best detected by several different instruments recording various measures simultaneously. Thus, identifying specific changes in the brain's blood flow, the brain's electrical activity, and the body's response such as changes in blood pressure and heart rate, may eventually be the best way to identify such states. For the purposes of our initial studies, we wanted to try to determine the person's own subjective sense of the experience. This is why we had the meditators incorporate the tugging of the string into their meditative practice so that it would give us a signal as to when they were achieving the deepest part of their meditation while not disturbing the meditation. Since we were studying highly proficient meditators, the use of the string was felt to be a minimal, if any, distraction to these meditators. Later studies will be needed to explore these states in

more detail. Suffice it to state here that while it is difficult to determine precisely when and how the peak state occurs, it is conceivable to study such a state or at least extrapolate from other "lesser" states. Two other investigators that had an important role in these studies were Dr. Abass Alavi, the Chief of Nuclear Medicine at the Hospital of the University of Pennsylvania, who has been wonderfully supportive of all of my somewhat unusual endeavors; and Dr. Michael Baime, an internist affiliated with the University of Pennsylvania, who is also a practicing Tibetan Buddhist.

2. The word "soul" here is used in its broadest term and is not meant to confuse Eastern and Western ideas about religion and spirituality. Buddhist beliefs are highly complex and difficult to understand within a Western framework. However, we have tried to simplify the concepts in this vignette.
3. The characteristics of the experience most commonly reported by our subjects included senses of unity with the world, dissolution of the self, and intense emotional responses usually associated with profound calmness.
4. Typically, science requires that something be measurable in order for it to be "real."
5. The term *real* here does not necessarily imply that there is an external reality associated with the experience, only that the experience is at least internally real.
6. While we realize that it is much more complex than simply "taking a picture," this is the gist of what is being done. Actually capturing the precise moment of an intense mystical experience is not easy, or likely, because in spite of the planned meditation our subjects performed, it is very difficult to know or predict exactly how long or how strong a given state will be. It, nonetheless, seems possible to begin to unravel the brain mechanisms that underlie the process of meditation and obtain

a clear view into the fantastic workings of the brain, during these practices.

7. There are several technologies similar to SPECT imaging that can be used to measure brain function. These include positron emission tomography (PET) and functional magnetic resonance imaging (fMRI). Each of these techniques have specific advantages and disadvantages compared to the other techniques. For our purposes, SPECT imaging allowed for the most realistic setting for practicing meditation because the subjects could perform their meditation outside of the scanner, something that is difficult with PET and impossible with fMRI.
8. In general, blood flow is associated with increased activity because the brain self-regulates the flow of blood to accommodate its need for additional energy. However, this is not always the case. Stroke and head trauma cause a detachment between blood flow and activity. Also, some nerve cells cause excitation in other parts of the brain while other nerve cells cause inhibition. Therefore, increased blood flow may reflect increasing inhibition resulting in an overall decrease in function.
9. We should note here that many of the terms utilized throughout this book are not specifically scientific, but rather are designed to make the complex workings of the brain easier to understand. However, we will try to indicate what the scientific terms are as a reference for those interested.
10. Throughout this book we will also refer to various functions of different parts of the brain. While it is possible to localize function to a reasonable degree, it is important to realize that the brain needs to work as a unit with each part requiring the other parts to work normally.
11. This type of blocking of input has been shown to occur in both normal and pathological states. Blocking of input can also occur in a variety of brain structures via various inhibitory

influences throughout the brain. We will consider this in more detail later.

12. Quoted in Easwaran 1987.
13. While not all subjects demonstrated a specific decrease in the activity in the orientation area, there was a strong inverse correlation between increasing activity in the frontal lobe (the area of the brain involved in focusing attention) and the orientation area. The implication of this finding is that the more the subjects were able to focus during their meditation, the more they were able to block input into the orientation area. The reason not all subjects were found to have a specific decrease in activity in the orientation area has two possible explanations. First, the subjects without a decrease may not have been able to meditate as successfully as the others, and while we tried to determine exactly how well their meditation went, it is a highly subjective state that is very difficult to measure. Second, we were limited in this study by looking only at one time point of the meditation. It is possible that in the early stage of meditation there is actually an increase in activity in the orientation area while the subject begins his focus on a visualized object. Thus, we might capture the orientation area either increased, unchanged, or decreased depending on the stage of meditation the person was actually in even though they may have described themselves as being in a deeper stage. We will discuss the implications of these findings in more detail in the chapter devoted to mystical experiences.
14. For a more detailed discussion of these experiments, see Newberg et al. 1997, 2000.
15. We will use the conventional naming of God in the male gender even though God can be envisioned in any manner.
16. These studies were only our initial empirical exploration of the neurophysiology of spiritual experience. However, our results, with those of other investigators (see Herzog et al.

1990–1991, Lou et al. 1999), supported some of the important points of our hypothesis.

Chapter 2: Brain Machinery

1. The other possibility, of course, is that a divine creator produced the wonderfully complex workings of the brain. Either way, organic brains are significantly more adept at processing sensory input from the outside world to produce a fluid version of reality. However, for the purposes of this book, we will use the scientific interpretation of evolutionary development.
2. The functioning of nerve cells across species naturally has some distinctions; however, the basic electrochemical processes are remarkably similar. In fact, much of our current understanding about the biology of the nerve cell arises from our study of other animals. Simpler animals such as worms and mollusks are easier to study because they only contain a few thousand nerve cells while in the human brain there are billions. Nerve cells do become differentiated in the brain so that they can play different roles and utilize different chemicals as neurotransmitters, the method of communication between nerve cells. For example, some nerve cells rely on a chemical called dopamine, others on acetylcholine, but they all respond to incoming stimuli and produce some form of output. Much of this is controlled by the movement of various ions such as sodium and calcium across the cell surface. Finally, there are other types of cells in the brain, particularly the myelin cells that are crucial to the transmission of signals between nerve cells. Disorders such as multiple sclerosis are the result of a destruction of these myelin cells making the individual's nerve cells unable to communicate with each other, resulting in a variety of neurological problems both cognitive and motor related.
3. Freshwater flat worms were possibly the earliest species to

have a central nervous system (Colbert 1980, Jarvic 1980). Furthermore, these nervous systems may have first evolved several hundred million years ago (Joseph 1993).

4. It should be made clear that the progression toward greater complexity is not necessarily linear. In other words, each subsequent animal does not necessarily have a larger and more complex brain. The overall trend, however, is that with progressive evolution there is a progressive increase in the size (with respect to body size) and complexity of the brain that allows for more flexible adaptability in an ever-changing environment. For a more complete description of brain evolution, see Joseph 1996.
5. There are a number of other structures that may eventually be found to be involved in spiritual experiences, such as the thalamus, the reticular activating system, the septal nucleus, and the cerebellum. These structures have not been extensively studied in terms of their relationship to spiritual experience, but because of the important roles they play in overall brain function, it's likely that future research may prove them to play a significant role and therefore we felt it important to at least mention briefly what these structures do. The thalamus is the major relay between different brain structures. It helps connect the frontal lobes to the limbic system, and connects much of the neocortex to the subcortical structures as well as the rest of the body. Because of these important connections, the thalamus plays a major role in complex brain functions. The reticular activating system (RAS) is a group of wispy nerve cells situated at the top of the spinal cord. The RAS helps to regulate arousal responses. It is likely that the RAS is involved with unusual hyperarousal states. The septal nucleus acts in conjunction with a structure called the hypothalamus and the hippocampus to exert quieting and dampening influences on arousal and limbic system function. This also means that the septal nucleus aids in selective attention and memory formation. The cerebellum, which sits near

the base of the brain toward the back, was long believed to be primarily involved in the motor system. Recent brain imaging studies have suggested that the cerebellum may also play a role in the control of attention, learning, and internal control of timing among other functions. At this time, however, the cerebellum's role in higher brain functions and especially religious experience is not completely clear.

6. Refer to either Kandel, Schwartz, and Jessell 2000 or Joseph 1996 for more technical details of the differences in function between the hemispheres.
7. The actual problem solving abilities of the brain are obviously much more complex, but this example demonstrates the fact that even when connected normally, the hemispheres do not share highly specific information but communicate in a more vague fashion.
8. The actual ability of these connecting structures to transmit information is quite complex. For example, while the main structure connecting the hemispheres, the corpus callosum, allows for the transmission of information, it also limits information (Selnes 1974). Thus, information may be lost or degraded during transfer (Marzi 1986). Even information that is transferred may be subject to misinterpretation by the receiving hemisphere (Joseph 1988, Gazzaniga and LeDoux 1978).
9. See Kandel, Schwartz, and Jessell 2000.
10. See Sperry 1966 for a historical description.
11. For additional reading on the disconnection syndrome and the connection aspects of the corpus callosum, see Gazzaniga, Bogen, and Sperry 1962 and Joseph 1988.
12. There may be as many as forty or more different types of neurons based on how they are structured and how they function. However, all nerve cells share a number of basic physiological functions that have to do with the transmission of information via electrochemical changes. For a detailed review

of neurons and their different functions, see Kandel, Schwartz, and Jessell 2000.

13. Most sensory systems have a specific neural structures for processing. These are called *unimodal* areas of the brain because they participate in only one type of sensory processing. Other parts of the brain combine input from different sensory systems and are therefore called multimodal. See Kandel, Schwartz, and Jessell 2000.
14. See Weiskrantz 1986, 1997.
15. See Gloor 1990 and Penfield and Perot 1963.
16. These are not the technical names for these association areas, but we will use these connotations since they better describe the functions and will help toward simplifying the neurophysiology. We will try to indicate where in the brain these specific areas are actually located.
17. A number of brain imaging studies using functional magnetic resonance imaging (fMRI) have demonstrated how this area is activated during tasks that require the manipulation of three-dimensional space as well as how this area functions along with the attention association area for spatial memory (Cohen et al. 1996, d'Esposito et al. 1998). For a further detailed discussion of the orientation association area, see Joseph 1996.
18. Rhawn Joseph (1996) was one of the first neuroscientists we are aware of to reach this conclusion. It should be mentioned, however, that to fully identify the self probably requires other structures, particularly those in the subcortical areas that are involved in the basic maintenance of the self. It is likely that as the human brain evolved, this function became incorporated and enhanced by the function of the OAA so that we can experience a rich sense of the self.
19. The prefrontal cortex actually has many different complex functions. However, for the purposes of this book, we will focus primarily on its ability to help us to focus attention.

20. In terms of the attention association area's function, an early study (Ingvar and Philipson 1977) examined blood flow in the brains of subjects moving their hand in a willful, rhythmic hand-clenching motion and also when they only imagined doing so. The findings showed increased activity in the motor area in the first condition and increased activity in the attention association area in the second condition. When imaging the movement, there was no increase in activity in the motor area. A PET study by Frith et al. (1991) showed that not only did willful acts cause an increase in the attention association area activity, but it also resulted in decreased activity in certain brain regions, including the posterior temporal-parietal area during a sensorimotor task. This was also found to be the case with other areas involved in modality-specific functions. It should be mentioned here that imaging studies measure certain biological functions but may not necessarily reflect accurately what is going on. For example, increased metabolic activity in inhibitory neurons will look the same on a PET scan as increased activity in excitatory neurons, but the cognitive results could be very different. The attention association area is also divided into various parts. And while we will not discuss the specific function of these various parts, it is important to keep in mind that the attention association area has many functions, some of which are very subtle. Many other imaging studies have also demonstrated similar findings, including those of Michael Posner and his colleagues (1990, 1994) and others (Pardo et al. 1991).
21. Rhawn Joseph is one of the first neuroscientists that we know of to make such a reference of the attention area as the seat of the will. However, there have clearly been other studies and investigators that have made a similar link (see Libet, Freeman, and Sutherland 1999, and Frith et al. 1991).
22. See Pribram and McGuinnes 1975 and Pribram 1981.
23. The frontal lobes have also been found to be highly involved in

emotions, and having intimate connections with the limbic system, the part of the brain most directly tied to emotions. The frontal lobes are believed to be extremely important in initiating, controlling, and monitoring emotions. Thus, it heavily interacts with the limbic system. People who have damage to the frontal lobes, like the classic case of Phineas Gage, generally begin to lose their personality and eventually have problems controlling their emotions or have blunted emotions (Damasio 1994). If there is extensive damage to the frontal lobes, the usual result is apathy, loss of emotional response, and loss of social interests (Damasio 1994; Kandel, Schwartz, and Jessell 2000). However, if only small areas are damaged, specifically those that control emotions, then the person can be disinhibited, hyperactive, euphoric, and emotionally labile. It is interesting to note that a PET study of murderers demonstrated decreased frontal-lobe glucose metabolism (Raine et al. 1994).

24. In this experiment (Ryding et al. 1996), counting out loud and counting silently were compared. Counting aloud demonstrated increased activity in the motor area involved in working the mouth, tongue, and lips. Counting silently resulted in increased activity in the attention association area. These studies were not done with the same high-resolution methods currently available, but demonstrate the importance of the attention association area in willful performance of different behaviors.
25. See Newberg et al. 1997, 2000 and Herzog et al. 1990–1991.
26. See Hirai 1974.
27. See Kandel, Schwartz, and Jessell 2000 for a detailed discussion of the verbal-conceptual area that represents the junction of the temporal and parietal lobes.
28. These patients were also reported to display an unusual obsession with religious matters, and even report feelings of union with the universe. Michael Persinger, from the Laurentian University in Ontario, has explored the relationship between temporal-lobe

seizures and religious experience (1993, 1997). He has shown that temporal lobe seizures and even electical stimulation can produce experiences similar to those described by meditators and people who have had a near-death experience. These include certain visions, emotional responses, and out-of-body experiences.

29. Dualism separates mind from brain as in Cartesian dualism. Damasio considers the relationship between the mind and brain in *Descartes' Error* (1994) suggesting a more materialistic perspective. Staunch materialists believe that mind is brain, or more specifically that the mind is completely derived from the function of the brain. While we do agree that the mind is derived from the brain, we also believe that the interaction is much more complex and intriguing. In *Brain, Symbol, and Experience* (Laughlin, McManus, and d'Aquili 1992) this relationship is defined as follows: "[our hypothesis] specifically holds that 'mind' and 'brain' are two views of the same reality—mind is how the brain experiences its own functioning, and brain provides the structure of mind." This is a slightly different view from the materialists and is also distinct from the dualists. However, we feel that this is the most accurate way of considering these concepts as we will describe them throughout this book. Throughout the remainder of this book, we will try to use the terms "mind" and "brain" according to our defintion, but as we have stated, they are two ways of describing the same thing.

Chapter 3: Brain Architecture

1. Even if there were a soul through which God could communicate, it would have little cognitive meaning to us without a brain.
2. There is a third component to the autonomic nervous system that regulates the gut, but for the purposes of discussing spiritual experiences, we will restrict our discussion to the sympathetic and parasympathetic systems.

3. For a detailed description of the general functioning of the autonomic nervous system see Kandel, Schwartz, and Jessell 2000.
4. For a more thorough discussion of the interaction between the different arms of the autonomic nervous system, see Hugdahl 1996. It is also interesting to note that just activation of one of the arms of the autonomic nervous system when stimulated by pharmacological means does not necessarily result in unusual emotional and cognitive states, although sometimes this can be the case. Thus, the states that we will be describing below may be expressed via activity in the autonomic nervous system, but these states are also generated in conjunction with other associated brain structures.
5. Essentially any repetitive stimulation, whether it be physical, emotional, sensory, or cognitive, can potentially generate such states (Gellhorn and Kiely 1972).
6. See Jevning et al. 1992, Corby et al. 1978, and MacLean et al. 1994 for details of the autonomic changes during meditation. While extrapolating meditative practices to all religious experiences is difficult because of the complexity of these experiences and the different types of autonomic responses that occur during them, throughout the remainder of this book we will expand upon the relationship between the autonomic nervous system, in addition to other brain structures, in order to better understand their relationship to religious experience.
7. For a more detailed description of these specific autonomic states, see Gellhorn and Kiely 1972, 1973, and Lex 1979.
8. We have previously described these states (see d'Aquili and Newberg 1999, Newberg and d'Aquili 2000). Specific evidence for these states, especially associated with spiritual practices, is not complete because the differential activity in the autonomic nervous system is difficult to measure. Typical measures include heart rate, blood pressure, and other physiological functions. The problem comes from interpreting the meaning of the

changes in these measures. For example, increased heart rate may be associated with either an increase in arousal or a decrease in the quiescent activity. Thus, it is very difficult to measure when both arms of the autonomic nervous system are stimulated at the same time. Some investigators have shown evidence of arousal and quiescent activity at the same time based on both body physiology measures as well as changes in brain activity. In other words, there is evidence of quiescent activity during meditation, but there is also some clear arousal activity since they maintain their attention and focus during the meditation. This has recently been demonstrated in a study of heart rates during meditation in which there was an increase in the oscillation between heart rates during the meditation period (Peng et al. 1999). This suggested that the autonomic activity is highly variable during such states and also indicates that autonomic activity during practices such as meditation may be quite complex.

9. See Czikszentmihalyi 1991.
10. See Weingarten et al. 1977, Horowitz et al. 1968, and Halgren et al. 1978.
11. See Lilly 1972, Zuckerman and Cohen 1964, and Shurley 1960.
12. Rhawn Joseph (2000) used this phrase in the book of the same title. He emphasized the importance of the limbic system in religious phenomena. Other researchers such as V. S. Ramachandran and Michael Persinger (1993, 1994) have singled out the limbic system as the major actor in religious experience. We would suggest, however, that the temporal lobe and the limbic structures within it cannot be solely responsible for the complexity and diversity of these experiences. We believe that there are many other structures that are involved in such experiences. In his book *Zen and the Brain*, James Austin described some of these structures associated with meditation, including those in the thalamus and temporal lobe. He also attempted to define some of the neurotransmitter systems that might be involved. We

feel that this may be somewhat premature since we have not fully described the particular structures that are involved and need to do this before we can attempt to uncover specific neurochemical processes. Regardless, his theory dovetails nicely with ours, although he appears less convinced about the significance of the autonomic nervous system in these states. We believe that all of these parts are important and that they are all required in order to explain the vast array of religious and spiritual experiences.

13. The hypothalamus is a relatively small structure but has a large number of functions crucial to our survival. In addition to its ability to regulate the arousal and quiescent systems, it helps to regulate aggression, sex, and certain survival-related behaviors. The hypothalamus regulates many hormonal systems in the body, including reproductive hormones, the thyroid, and growth hormone, and it also moderates immune function, hunger, thirst, and body temperature. Thus, the hypothalamus is incredibly important and necessary for a wide array of brain functions. For more details on the functioning of the hypothalamus, see Kandel, Schwartz, and Jessell 2000.
14. See MacLean et al. 1994.
15. For a discussion of the neuroimaging studies demonstrating the function of the amygdala as well as a detailed neurophysiological review of the relationship between the amygdala and emotion, motivation, and attention, refer to Gazzaniga 2000.
16. See Halgren 1992.
17. For more details of the functioning of the amygdala and hypothalamus in the arousal response, see Kandel, Schwartz, and Jessell 2000.
18. For more details on how the hippocampus links emotions to images and memory, see Joseph 1996.
19. For more details on how the hippocampus interacts with the thalamus in order to block sensory input, see either Kandel, Schwartz, and Jessell 2000 or Joseph 1996.

20. The specific concept of a cognitive operator was previously developed in our work as a way of describing general functions of the brain. These clearly are similar to the concept of cognitive modules in that both are functions and are localizable to one or more specific areas of the brain. However, we will continue to utilize the term *cognitive operator* since we are referring to general ways in which the brain operates on various sensory or cognitive input. It should be mentioned that most of the evidence that demonstrates the use of the concept of cognitive modules also applies to the cognitive operators. We will present evidence suggesting that the cognitive operators presented in this book are, in fact, specific ways in which the brain processes information and that these ways are an inherent part of the functioning of the brain. Hence, it can also be argued that there are good biological as well as evolutionary reasons why cognitive operators (or modules) exist. Martin, Ungerleider, and Haxby (2000) state that there are prelanguage ways for "processing and storing information about form, color, motion, and movement," and that a good case can be made with regards to the processing of "space, time, number, and affective valence." We will see how the idea of cognitive operators fits this description.
21. Cognitive operators differ from *cognitive modules*, a term described by neuroscientists such as Steven Pinker (1999). We feel that the cognitive modules represent more specific functions that are localized to particular brain structures. For example, a module associated with mathematics would refer only to one brain function such as basic arithmetic, whereas the quantitative operator refers to the many mathematically related functions of the brain all at once.
22. Evidence for the holistic operator derives from studies that have explored the functions of the right hemisphere demonstrating more holistic applications to perceptions and problem solving (Schiavetto, Cortese, and Alain 1999; Sperry, Gazzaniga,

and Bogen 1969; Nebes and Sperry 1971; Gazzaniga and Hillyard 1971; Bogen 1969). The reductionist operator clearly is associated with our deductive abilities usually ascribed to the left temporal-parietal region (Luria 1966; Basso 1973). The abstractive operator likely resides in the region of the left inferior parietal lobe, most likely near the angular gyrus, and forms an important part of the language axis (Luria 1966, Geschwind 1965; see also Joseph 1996). The origin of the quantitative operator is somewhat more complex than we have previously stated because it appears that both the left and right inferior parietal areas may be involved. The left hemisphere is more associated with specific mathematical functions while the right appears better equipped for comparing numbers. For a thorough discussion of the brain's quantitative abilities, see Dehaene 2000, in which he states that "studies of number production, comprehension, and calculation tasks have provided strong evidence for a modular organization." Our concept of the cognitive operator is merely used as a more global term even though there may be more specific modules by which we handle quantitation, see Pesenti et al. 2000. The causal operator has much scientific support, although from slightly older studies (Pribram and Luria 1973, Mills and Rollman 1980, Swisher and Hirsch 1971). A recent study (Wolford 2000) demonstrated that human beings seek out sequences of events even when they are told that the events are random. Further, this function was shown to be primarily a left hemispheric function. The binary operator appears to have a conceptual basis (Murphy and Andrew 1993) and may arise near the quantitative operator in the region of the left inferior parietal lobe (Gardner et al. 1978, Gazzaniga and Miller 1989). This area also appears to be able to distinguish concepts of "greater than" or "less than" with regards to numbers (Dehaene 2000). The existential operator has not been previously described, but appears to be an important aspect of human perception. The very fact that we believe that objects that we perceive do in fact exist is perhaps one of the most fundamental functions of the brain, but clearly can go awry in terms of hallucinations, or utilized to fool us by a magician. In terms of the emotional value operator, much evidence for the importance of emotions in human behavior and reasoning has come from the research of Antonio Damasio (1994, 1999). His somatic marker hypothesis suggests that emotions are critical in helping human beings make decisions and think rationally. We would agree that what we are calling the emotional value operator is of critical importance in terms of how we order and relate to the world. Furthermore, emotions appear necessary to assign relative value to all of the other products of the cognitive operators. With regards to religious experience, some have indicated that the limbic system is the "seat of the soul" because it provides the emotional value, or strength, to various experiences and marks them as spiritual (Joseph 2000, Saver and Rabin 1997). Our model would also support this contention in that the limbic system is critical in valuing certain experiences and perhaps even participating in the evaluation of the reality of those experiences. This will become crucial later on. For now it is enough to state that the limbic system, which underlies the function of the emotional value operator is clearly crucial for helping us order and respond to our world. It has a role not only in "irrational" ideas but in rational ones as well.

23. For a discussion of the studies exploring infants' ability to perform arithmetic, see Bryant 1992.
24. For a thorough discussion of the experiment, see Spelke et al. 1992.
25. See Damasio 1999.

Chapter 4: Myth-making

1. There is abundant evidence of religious activities, or at least a belief in life after death, from findings associated with mortuary practices (Belfer-Cohen and Hovers 1992, Butzer 1982, Rightmire 1984, Smirnov 1989). These findings are scattered throughout Europe and Africa and demonstrate widespread use of such practices. For a more detailed discussion of these practices, see Joseph 2000.
2. See Kurten 1976 and Joseph 2000.
3. Quoted in Campbell 1972.
4. Ibid.
5. It is particularly interesting to note that higher order thought regarding fearful situations undoubtedly requires memory of past traumatic events in association with limbic activation during the current situation. Interestingly, both memory and limbic function are mediated in large part by the amygdala and often in conjunction with the hippocampus. For a more detailed discussion see Damasio 1999, Joseph 1996, and LeDoux 1996.
6. Not only is social attachment adaptive, but it is ingrained in our brains as demonstrated by the strong need for contact that all infants have. This includes not only for their mother but for all people with whom the infant comes in contact. Furthermore, animals raised in social isolation will seek out social contact with inanimate objects or even predators (see Harlow 1962, Cairns 1967).
7. As with the cognitive operators, the cognitive imperative is not necessarily a thing in itself but refers to the brain's function to order our world in an almost automatic way. In other words, the brain is not like a computer that can be completely turned off. It is always doing something, even while asleep.
8. The concept of the cognitive imperative was originally used by Eugene d'Aquili in 1972. There are a number of hallmarks of

the cognitive imperative. These include the frustration of the imperative in the face of unyielding novel input, which has been shown to lead to anxiety. In fact, studies have shown that the brains of higher organisms tend to strive for a balance between novelty and redundancy (Berlyne 1960, Suedfeld 1964). Too much novelty is met with attempts to classify the input into simpler categories. On the other hand, too little novelty causes the brain to become bored and forces it to create uncertainty or complexity. We also see evidence for the cognitive imperative in what a group of scientists described as ontological yearning, that desire to understand the fundamental nature of the world (see Larson, Swyers, and McCullough 1997). Anthropologist Misia Landau (1984) also states that to overcome the anxiety derived from the cognitive imperative, we have "basic stories, or deep structures, for organizing our experiences." Finally, E. O. Wilson (quoted in Shermer 2000) describes how storytelling brings into "play all of the cognitive and emotional circuitry evolved to deal with real experience." Thus, with the functioning of the cognitive imperative, we cannot help but to organize the world and our experience of the world by creating stories and ultimately myths to help perform this function.

9. For a more detailed description of the mythic framework, see d'Aquili 1978, 1983, and d'Aquili and Newberg 1999.
10. For a more detailed discussion of the evolution of the human brain and its distinguishing features compared to other primate species, see Preuss 1993, 2000.
11. The use of endocasts is the only way to evaluate the structure of early hominid brains (see LeGros Clark 1947, 1964, and Holloway 1972). The curves and imprint of the structures on the inner surface of the skull can be used to infer the structures that lie adjacent to them. There are, however, other

investigators who argue against the use of such methods to determine what the brain actually looked like or even what functions it was capable of performing (Jerison 1990).

12. There does not appear to be a structure in other primates or in hominid ancestors of a degree and complexity to the inferior parietal area (Wernicke's area) in human beings that would be capable of sustaining language and verbal speech. This conclusion is based on anatomical studies of current primate species as well as endocasts of human ancestors (Holloway 1972, Joseph 1993). It should be mentioned that there are structures in certain primate species such as the macaques that share similarities to the human parietal area (Galaburda and Pandya 1982). However, these areas in macaques do not have the same degree of complexity, nor the interconnections with other areas that would enable them to sustain language.
13. During the latter stages of *Homo erectus* development, about five hundred thousand years ago, there was a significant increase in brain size (Rightmire 1990).
14. See d'Aquili, Laughlin, and McManus 1979.
15. The problem-solving process that we describe here is obviously simplified. Still, we feel it is congruent with current research on human problem solving. Also, the notion that the right brain utilizes visceral, autonomic, and somatic information to help solve an analytic problem is consistent with Damasio's somatic marker hypothesis.
16. For an excellent review and discussion of Jung's ideas about myth and archetypal constructs, see Jung 1958.
17. See *Myths to Live By* (Campbell 1972).

Chapter 5: Ritual

1. For a detailed description of how music can affect the brain, see Iwanaga and Tsukamoto 1997.

2. Drumming rhythms vary in a given performance to evoke responses in most listeners so that individual differences in basal rhythms are accommodated (Neher 1962). Thus, each individual gets something out of the specific rhythms that resonates with them. This makes rhythmic group rituals more globally effective. Body movement during ritual stimulates proprioceptors resulting in dizziness and disturbances in the vestibular system that maintains balance and equilibrium. Repetitive muscle tension and relaxation can also effect emotional responses (Gellhorn and Kiely 1972, 1973). Neher also showed that there are multiple sensory modalities involved in the rhythms and there are also contributing factors such as fasting, hyperventilation, and various smells that can all affect the body's physiology. Other investigators (Walter and Walter 1949) have also shown that repetitive auditory and visual stimuli drive cortical rhythms and can produce intensely pleasurable, ineffable experiences in human beings. Repetitive stimuli can also evoke intense discharges from both the arousal and quiescent systems. A recent study demonstrated that during meditative practices, there was a significant variation in heart rate indicating that there are significant alterations in the autonomic nervous system and not just a relaxing response (Peng 1999).
3. See d'Aquili, Laughlin, and McManus 1979.
4. Burns and Laughlin (1979) provide an ample review of studies demonstrating how ritual functions as a mechanism of social control, resolves social conflict, maintains social solidarity and social stratification, and maintains the power structure of a given society. See also Turner 1969 and Blazer 1998.
5. See d'Aquili, Laughlin, and McManus 1979.
6. Ibid.
7. See d'Aquili and Newberg 1999 and d'Aquili 1983.
8. See Smith 1979.
9. See Bastock 1967.

10. See Smith 1979.
11. See d'Aquili, Laughlin, and McManus 1979.
12. See Smith 1979.
13. See d'Aquili and Newberg 1999 and d'Aquili 1983.
14. See d'Aquili and Newberg 1999. It is interesting to note that some nomadic societies do not have transcendence as a specific goal of their ritual (Berman 2000). However, this may be due to the fact that these groups already feel connected to the oneness of the world. Cultures that do not feel such connectedness often use ritual for this purpose.
15. See Larsen, Swyers, and McCullough 1997; Koenig 1999; Corby 1978; and Jevning 1992.
16. We have described earlier how the hippocampus works to prevent extremes of function in various parts of the brain, particularly in the limbic system.
17. Many of the people who have participated in our study use phrases like "powerfully calming," "ecstatic," "extremely peaceful," and "joy" and even described occasional negative emotions such as fear or anger. It should be mentioned that many others have used powerful emotional phrases to describe the effects of ritual and meditation.
18. See Gellhorn and Kiely 1972, 1973.
19. See d'Aquili and Newberg 1993.
20. For a detailed description of olfaction and the olfactory cortex, see Kandel, Schwartz, and Jessell 2000.
21. For details of this study, see Vernet-Maury et al. 1999.
22. See Gellhorn and Kiely 1972, 1973.
23. See Joseph 1996, Savitzky 1999, Collet et al. 1997, Smith et al. 1995, and for parasympathetic effects, Porges et al. 1994.
24. See Secknus and Marwick 1997.
25. This concept may appear somewhat similar to the somatic marker hypothesis of Antonio Damasio we described earlier. His theory,

however, emphasizes the input from the various sensory organs. We agree with this analysis, but also feel that autonomic function is an important component.

26. See Telles, Nagarathna, and Nagendra 1995, 1998.
27. Quoted in Campbell 1988.
28. Quoted in Campbell 1968.
29. See Lajonchere, Nortz, and Finger 1996.
30. Ibid.
31. For a discussion of echolalia and its relationship with frontal lobe dysfunction, see Hadano, Nakamura, and Hamanaka 1998 and Vercelletto et al. 1999.
32. A number of studies have demonstrated the importance of visualizing our behaviors and actions, including the use of brain imaging studies (see Jeannerod and Frak 1999 and Lotze et al. 1999). The results indicate that imagining actions is adaptive and utilizes similar, but different, parts of the brain than those used only for generating actions.

Chapter 6: Mysticism

1. Quoted in Cooper 1992.
2. For a very thorough discussion of how Freud considered religion and the basis for his theory of religion, see Kung 1990.
3. See Underhill 1990.
4. Quoted in Underhill 1990.
5. Quoted in Nicholson 1963.
6. Quoted in Underhill 1999.
7. Quoted in Kabat-Zinn 1994.
8. Ibid.
9. Quoted in Teasdale 1999.
10. Quoted in Epstein 1988.
11. Ibid.

12. See Hodgson 1974.
13. See James 1963.
14. See Greeley 1987. Follow-up studies have also yielded similar results.
15. See Saver and Rabin 1997.
16. For research regarding the improved mental health and changed perspective of people who have had near-death experiences, see Greyson 1993, Bates and Stanley 1985, Noyes 1980, and Ring 1980. For general religious experiences, see Koenig 1999.
17. See Lilly 1972, Shurley 1960, and Zuckerman and Cohen 1964.
18. Quoted in Underhill 1990.
19. We have previously described the difference between fast and slow rituals and how they affect the activity in the arousal and quiescent arms of the autonomic nervous system (see d'Aquili and Newberg 1999).
20. These two approaches are meant to be extremely broad and inclusive. There are, of course, thousands of meditative approaches, some of which have a combination of both types of techniques. However, for the practical purposes of trying to devise a neurological model of such states, it is necessary to find some basic structural aspects of different types of meditation. The bottom line is that this model is attempting to lay a general foundation upon which we can begin to discuss how a large variety of practices may result in widely different experiences. Since there are clearly certain similarities between experiences, we believe that different practices, including spontaneous experiences, likely activate similar pathways although in slightly different ways. These differences are what give particular practices their differences and resulting experiences. We first described this model in detail in our 1993 article in *Zygon* and further elaborated that model in our 1999 book.
21. A number of electroencephalographic (EEG) studies have demonstrated increased electrical activity over the frontal lobes dur-

ing various types of meditation (see Benson et al. 1990, Anand et al. 1961, Banquet 1972).

22. While this deafferentation is known to occur in many circumstances, it has not been completely proven to occur during meditation practices. However, our study of Tibetan Buddhist meditators and another brain imaging study of yoga relaxation meditation have demonstrated relative increases in the frontal lobes and relative decreases in the posterior parietal lobes (see Newberg et al. 1997, 2000 and Herzog et al. 1990–91).
23. Many studies of various meditative techniques have demonstrated both times of increased and decreased autonomic activity (see Jevning et al. 1992, Benson et al. 1990, Peng et al. 1999, Sudsuang et al. 1991).
24. See Komisaruk and Whipple 1998, Knobil and Neill 1994.

Chapter 7: The Origins of Religion

1. For a thorough discussion of the psychological benefits associated with religiousness see Koenig 1999 and also Worthington, Kurusu, McCullough, Sandage 1996. These researchers discuss a number of possible mechanisms such as increased social support and increased sense of meaning in life.
2. See the American Psychiatric Association DSM-IV 1994.
3. See Koenig 1999 and also Worthington, Kurusu, McCullough, and Sandage 1996.
4. While there are numerous studies on the relationship between social support and religious practice, several specific articles can be referred to (see Krause et al. 1999 and Oman and Reed 1998).
5. See Jevning, Wallace, and Beidebach 1992 and Kesterson 1989.
6. See the NIHR consensus report and spirituality and health (Larson, Swyers, and McCullough 1997).
7. See Teasdale 1999.

Chapter 8: Realer Than Real

1. Quoted in Armstrong 1993.
2. Quoted in Blofield 1970.
3. Quoted in Kabat-Zinn 1994.
4. For a more detailed discussion of the comparison between material reality and subjective reality, see d'Aquili 1982 and Newberg 1996.
5. Quoted in Hoffman 1981.
6. Quoted in Schrödinger 1964.
7. Quoted in Reagan 1999.
8. Quoted in Sagan 1986.
9. The argument regarding Absolute Unitary Being as primary in the universe with subjective and objective deriving from it is a complex philosophical discussion. While we cannot at this time prove the notion that Absolute Unitary Being is, in fact, primary, with subjective and objective reality deriving from it, based on our phenomenological analysis of the different states of reality it seems a reasonable conclusion. Also, the existence of an all-encompassing, creative, transcendent reality would easily solve the problems objective and subjective realities cannot fathom: why we have consciousness, for example, or why the universe bothers to exist at all. As absolute, undifferentiated oneness, Absolute Unitary Being would resolve all existential questions and resolve the dilemma of opposites—life and death, good and evil, spirit and flesh, gods and humans—that compel us to make myths, and are the focus of all our spiritual strivings.
10. Quoted in Kabat-Zinn 1994.

Chapter 9: Why God Won't Go Away

1. Quoted from the Hindu scripture *Vasistha's Yoga.*
2. See al-Adawiyya 1988.

3. See Davies 1994.
4. Quoted in Underhill 1990.
5. See Teasdale 1999.
6. See Armstrong 1993.
7. Quoted by the Dali Lama.
8. See Teasdale 1999.
9. Quoted in Zukav 1979.

REFERENCES

Aggleton, J. P., ed. 1992. *The Amygdala.* New York: Wiley-Liss.

al-Adawiyya, R. 1988. *Doorkeeper of the Heart: Versions of Rabia,* tran. Charles Upton. Putney, Vt.: Threshold Books.

American Psychiatric Association. 1994. *Diagnostic and Statistical Manual of Mental Disorders: DSM-IV, 4th ed.* Washington, D.C. : American Psychiatric Association.

Anand, B. K., G. S. China, and B. Singh. 1961. Some aspects of electroencephalographic studies on Yogis. *Electroencephalography and Clinical Neurophysiology* 13:452–56.

Angela of Foligno. 1993. *Complete Works,* tran. Paul Lachance. Mahwah, NJ: Paulist Press.

Armstrong, E., and D. Faulk, eds. 1982. *Primate Brain Evolution.* New York: Plenum Press.

Armstrong, K. 1993. *A History of God.* New York: Ballantine Books.

Austin, J. 1998. *Zen and the Brain: Toward an Understanding of Meditation and Consciousness.* Cambridge, Mass.: MIT Press.

Banquet, J. P. 1972. EEG and meditation. *Electroencephalography and Clinical Neurophysiology* 33:454.

Basso, A., et al. 1973. Neuropsychological evidence for the existence of cerebral areas critical to the performance of intelligence tasks. *Brain* 96:715–728.

Bastock, M. 1967. *Courtship: An Ethological Study.* Chicago: Aldine Press.

Bates, B. C., and A. Stanley. 1985. The epidemiology and differential diagnosis of near-death experience. *American Journal of Orthopsychiatry* 55:542–549.

Belfer-Cohen, A., and E. Hovers. 1992. In the eye of the beholder: mousterian and Natufian burials in the levant. *Current Anthropology* 133:463–471.

Benson, H., M. S. Malhotra, R. F. Goldman, G. D. Jacobs, and P. J. Hopkins. 1990. Three case reports of the metabolic and electroencephalographic changes during advanced Buddhist meditation techniques. *Behavioral Medicine* 16:90–95.

Berlyne, D. 1960. *Conflict, Arousal, and Curiosity.* New York: McGraw-Hill.

Berman, M. 2000. *Wandering God: A Study in Nomadic Spirituality.* Albany, N.Y.: State University of New York Press.

Blazer, D. G. 1998. Religion and academia in mental health. In *Handbook of Religion and Mental Health*, ed. Koenig. San Diego: Academic Press.

Blofield, J. 1970. *The Zen Teaching of Huang Po.* New York: Grove.

Bogen, J. E. 1969. The other side of the brain. II: An appositional mind. *Bulletin of Los Angeles Neurological Society* 34:135–162.

Bryant, P. E. 1992. Arithmetic in the cradle. *Nature* 358:712–713.

Burns, T., and C. D. Laughlin. 1979. Ritual and social power. In *The Spectrum of Ritual*, eds. d'Aquili, Laughlin, and McManus. New York: Columbia University Press.

Butzer, K. 1982. Geomorphology and sediment stratiagraphy. In *The Middle Stone Age at Klasier River Mouth in South Africa*, eds. Singer and Wymer. Chicago: University of Chicago Press.

Cairns, R.B. 1967. The attachment behavior of mammals. *Psychological Review* 73:409–426.

Campbell, J. 1968. *The Masks of God: Creative Mythology.* New York: Viking/Penguin.

———. 1972. *Myths to Live By.* New York: Viking Press.

———. 1988. *The Power of Myth.* New York: Doubleday Books.

Cohen, M. S., S. M. Kosslyn, H. C. Breiter et al. 1996. Changes in cortical activity during mental rotation: A mapping study using functional MRI. *Brain* 119:89–100.

Colbert, E. H. 1980. *Evolution of Vertebrates.* New York: John Wiley & Sons.

Collet, C., E. Vernet-Maury, G. Delhomme, and A. Dittmar. 1997. Autonomic nervous system response patterns specificity to basic emotions. *Journal of the Autonomic Nervous System* 62:45–57.

Cooper, D. A. 1992. *Silence, Simplicity, and Solitude.* New York: Bell Tower.

Corby, J. C., W. T. Roth, V. P. Zarcone, and B. S. Kopell. 1978. Psychophysiological correlates of the practice of Tantric Yoga meditation. *Archives of General Psychiatry* 35:571–577.

Czikszentmihalyi, M. 1991. *Flow: The Psychology of Optimal Experience.* New York: HarperCollins.

d'Aquili, E. G. 1972. *The Biopsychological Determinants of Culture.* Massachusetts: Addison-Wesley Modular Publications.

———. 1975. The biopsychological determinants of religious ritual behavior. *Zygon* 10:32–58.

———. 1978. The neurobiological bases of myth and concepts of deity. *Zygon* 13:257–275.

———. 1982. Senses of reality in science and religion. *Zygon* 17:361–384.

———. 1983. The myth-ritual complex: A biogenetic structural analysis. *Zygon* 18:247–269.

———. 1985. Human ceremonial ritual and the modulation of aggression. *Zygon,* 20:21–30.

d'Aquili, E. G., and A. B. Newberg. 1993. Liminality, trance, and unitary states in ritual and meditation. *Studia Liturgica* 23:2–34.

———. 1993. Religious and mystical states: A neuropsychological model. *Zygon* 28:177–200.

———. 1996. Consciousness and the machine. *Zygon* 31:235–252.

———. 1999. *The Mystical Mind: Probing the Biology of Religious Experience.* Minneapolis: Fortress Press.

———. 2000. The neuropsychology of aesthetic, spiritual and mystical states. *Zygon* 35:39–52.

d'Aquili, E. G., C. Laughlin, Jr., and J. McManus, eds. 1979. *The Spectrum of Ritual: A Biogenetic Structural Analysis.* New York: Columbia University Press.

Damasio, A. R. 1994. *Descartes' Error: Emotion, Reason, and the Human Brain.* New York: Avon Books.

———. 1999. *The Feeling of What Happens: Body and Emotion in the Making of Consciousness.* New York: Harcourt Brace & Company.

Davies, O., trans. 1994. *Meister Eckhart: Selected Writings.* New York: Penguin Books USA, Inc.

Dehaene, S. 2000. Cerebral basis of number processing and calculation. In *The New Cognitive Neurosciences*, ed. Gazzaniga. Cambridge, Mass: MIT Press.

D'Esposito, M., G. K. Aguirre, E. Zarahn, D. Ballard, R. K. Shin, and J. Lease. 1998. Functional MRI studies of spatial and nonspatial working memory. *Cognitive Brain Research* 7:1–13.

Easwaran, E. ed. 1987. *The Upanishads.* Tomales, Calif.: Nilgiri Press.

Eccles, J. C., ed. 1966. *Brain and Conscious Experience.* New York: Springer Verlag.

Epstein, P. 1988. *Kabbalah: The Way of the Jewish Mystic.* Boston: Shambhala Publications, Inc.

Filskov, S. K., and T. J. Boll, eds. 1981. *Handbook of Clinical Neuropsychology.* New York: Wiley.

Frith, C. D., K. Friston, P. F. Liddle, and R. S. J. Frackowiak. 1991. Willed action and the prefrontal cortex in man. A study with PET. *Proceedings of the Royal Society of London* 244:241–246.

Galaburda, A. M., and D. N. Pandya. 1982. Role of architectonics and connections in the study of brain evolution. In *Primate Brain Evolution*, eds. Armstrong and Faulk. New York: Plenum Press.

Gardner, H., J. Silverman, W. Wapner, and E. Surif. 1978. The appreciation of antonymic contrasts in aphasia. *Brain and Language* 6:301–317.

Gazzaniga, M. S., J. E. Bogen, and R. W. Sperry. 1962. Some functional effects of sectioning the cerebral commissures in man. *Proceedings of the National Academy of Sciences* U8:1765–1769.

Gazzaniga, M. S., and S. A. Hillyard. 1971. Language and speech capacity of the right hemisphere. *Neuropsychologia* 9:273–280.

Gazzaniga, M. S., and J. E. LeDoux. 1978. *The Integrated Mind.* New York: Plenum Press.

Gazzaniga, M. S., and G. A. Miller. 1989. The recognition of antonymy by a language-enriched right hemisphere. *Journal of Cognitive Neuroscience* 1:187–193.

Gazzaniga, M. S., ed. 2000. *The New Cognitive Neurosciences.* Cambridge, Mass.: MIT Press.

Gellhorn, E., and W. F. Kiely. 1972. Mystical states of consciousness: Neurophysiological and clinical aspects. *Journal of Nervous and Mental Disease* 154:399–405.

———. 1973. Autonomic nervous system in psychiatric disorder. In *Biological Psychiatry*, Mendels. New York: John Wiley & Sons.

Geschwind, N. 1965. Disconnexion syndromes in animals and man. *Brain* 88:585–644.

Gloor, P. 1990. Experiential phenomena of temporal lobe epilepsy. *Brain* 113:1673–1694.

Greeley, A. 1987. Mysticism goes mainstream. *American Health* 6:47–49.

Greyson, B. 1993. Varieties of near-death experience. *Psychiatry* 56:390–399.

Hadano, K., H. Nakamura, and T. Hamanaka. 1998. Effortful echolalia. *Cortex* 34:67–82.

Halgren, E. 1992. Emotional neurophysiology of the amygdala within the context of human cognition. In *The Amygdala*, ed. Aggleton. New York: Wiley-Liss.

Halgren, E., T. L. Babb, and P. H. Crandall. 1978. Activity of human hippocampal formation and amygdala neurons during memory tests. *Electroencephalography and Clinical Neurophysiology* 45:585–601.

Harlow, H. F. 1962. The heterosexual affectional system in monkeys. *American Psychologist* 17:1–9.

Herzog, H., V. R. Lele, T. Kuwert, K. J. Langen, E. R. Kops, and L. E. Feinendegen. 1990–91. Changed pattern of regional glucose metabolism during Yoga meditative relaxation. *Neuropsychobiology* 23:182–187.

Hirai, T. 1974. *Psychophysiology of Zen.* Tokyo: Igaku Shoin.

Hodgson, M. G. S. 1974. *The Venture of Islam, Conscience and History in a World Civilization.* Chicago: University of Chicago Press.

Hoffman, E. 1981. *The Way of the Splendor.* Boulder, Colo.: Shambhala Publications, Inc.

Holloway, R. L. 1972. Australopithecine endocasts, brain evolution in the Hominoidea, and a model of hominid evolution. In *The Functional and Evolutionary Biology of Primates*, ed. Tuttle. Chicago: Aldine.

Hoppe, K. D. 1977. Split brains and psychoanalysis. *Psychoanalytic Quarterly* 46:220–244.

Horowitz, M. J., J. E. Adams, and B. B. Rutkin. 1968. Visual imagery on brain stimulation. *Archives of General Psychiatry* 19:469–486.

Hugdahl, K. 1996. Cognitive influences on human autonomic nervous system function. *Current Opinion in Neurobiology* 6:252–258.

Ingvar, D. H., and L. Philipson. 1977. Distribution of cerebral blood flow in the dominant hemisphere during motor ideation and motor performance. *Annals of Neurology* 2:230–237.

Iwanaga, M., and M. Tsukamoto. 1997. Effects of excitative and sedative music on subjective and physiological relaxation. *Perceptual and Motor Skills* 85:287–296.

James, W. [1890] 1963. *Varieties of Religious Experience.* New York: University Books.

Jarvic, E. 1980. *Basic Structure and Evolution of Vertebrates.* Vol. 2. New York: Academic Press.

Jeannerod, M., and V. Frak. 1999. Mental imaging of motor activity in humans. *Current Opinion in Neurobiology* 9:735–739.

Jerison, H. J. 1990. Fossil evidence on the evolution of neocortex. In *Cerebral Cortex*, eds. Jones and Peters. New York: Plenum.

Jevning, R., R. Anand, M. Biedebach, and G. Fernando. 1996. Effects of regional cerebral blood flow of transcendental meditation. *Physiology and Behavior* 59:399–402.

Jevning, R., R. K. Wallace, and M. Beidebach. 1992. The physiology of meditation: A review. A wakeful hypometabolic integrated response. *Neuroscience and Biobehavioral Reviews* 16:415–424.

Johnson, C. P. and M. A. Persinger. 1994. The sensed presence may be facilitated by interhemispheric intercalation: relative efficiency of the Mind's Eye, Hemi-Sync Tape, and bilateral temporal magnetic field stimulation. *Perceptual and Motor Skills* 79:351-354.

Jones, E. D., and A. Peters, eds. 1990. *Cerebral Cortex: Comparative Structure and Evolution of Cerebral Cortex.* Vol. 8B. New York: Plenum.

Joseph, R. 1988a. The right cerebral hemisphere: emotion, music, visual-spatial skills, body image, dreams, and awareness. *Journal of Clinical Psychology* 44:630–673.

———. 1988b. Dual mental functioning in a split brain patient. *Journal of Clinical Psychology* 44:770–779.

———. 1992. *The Right Brain and the Unconscious.* New York: Plenum.

———. 1993. *The Naked Neuron: Evolution and the Languages of the Body and Brain.* New York: Plenum.

———. 1996. *Neuropsychiatry, Neuropsychology, and Clinical Neuroscience.* Baltimore: Williams & Wilkins.

———. 2000. *The Transmitter to God: The Limbic System, the Soul, and Spirituality.* San Jose: University Press California.

Jung, C. G. 1958. *Psyche and Symbol*, tran. V. S. Laszlo. New York: Doubleday Anchor Books.

Kabat-Zinn, J. 1994. *Wherever You Go, There You Are: Mindfulness Meditation in Everyday Life.* New York: Hyperion.

Kandel, E. R., J. H. Schwartz, and T. M. Jessell. 2000. *Principles of Neural Science.* 4th ed. New York: McGraw Hill.

Kesterson, J. 1989. Metabolic rate, respiratory exchange ratio and apnea during meditation. *American Journal of Physiology* R256:632–638.

Knobil, E., and J. D. Neill, ed. 1994. *The Physiologist of Reproduction.* New York: Raven Press.

Koenig, H. G., 1999. *The Healing Power of Faith.* New York: Simon & Schuster.

———. ed. 1998. *Handbook of Religion and Mental Health.* San Diego: Academic Press.

Komisaruk, B. R. and B. Whipple. 1998. Love as sensory stimulation: physiological consequences of deprivation and expression. *Psychoneuroendocrinology* 23:927–944.

Krause, N., B. Ingersoll-Dayton, J. Liang, and H. Sugisawa. 1999. Religion, social support, and health among the Japanese elderly. *Journal of Health & Social Behavior.* 40:405–21.

Kung, H. 1990. *Freud and the Problem of God.* New Haven: Yale University Press.

Kurten, B. 1976. *The Cave Bear Story.* New York: Columbia University Press.

Lajonchere, C., M. Nortz, and S. Finger. 1996. Guilles de la Tourette and the discovery of Tourette syndrome. Includes a translation of his 1884 article. *Archives of Neurology* 53:567–574.

Landau, M. 1984. Human evolution as narrative. *American Scientist* 72:262–268.

Larson, D. B., J. P. Swyers, and M. E. McCullough. 1997. *Scientific Research on Spirituality and Health: A Consensus Report.* Rockville, Md.: National Institute of Healthcare Research.

Laughlin, C. Jr., and E. G. d'Aquili. 1974. *Biogenetic Structuralism.* New York: Columbia University Press.

Laughlin, C. Jr., J. McManus, and E. G. d'Aquili. 1992. *Brain, Symbol, and Experience,* 2d ed. New York: Columbia University Press.

LeDoux, J. 1996. *The Emotional Brain: The Mysterious Underpinnings of Emotional Life.* New York: Simon & Schuster.

LeGros Clark, W. E. 1947. Observations on the anatomy of the fossil Australopithecinae. *Journal of Anatomy* 81:300.

———. 1964. *The Fossil Evidence for Human Evolution.* Chicago: University of Chicago Press.

Lex, B. W. 1979. The neurobiology of ritual trance. In *The Spectrum of Ritual,* eds. d'Aquili, Lauglin, and McManus. New York: Columbia University Press.

Libet, B., A. Freeman, and K. Sutherland, eds. 1999. *The Volitional Brain: Toward a Neuroscience of Free Will.* Thorverton England: Imprint Academic.

Lilly, J. C. 1972. *The Center of the Cyclone.* New York: Julian Press.

Lotze, M., P. Montoya, M. Erb, et al. 1999. Activation of cortical and cerebellar motor areas during executed and imagined hand movements: An fMRI study. *Journal of Cognitive Neuroscience* 11:491–501.

Lou, H. C., T. W. Kjaer, L. Friberg, G. Wildschiodtz, S. Holm, and M. Nowak. 1999. A 15O-H2O PET study of meditation and the resting state of normal consciousness. *Human Brain Mapping* 7:98–105.

Luria, A. R. 1966. *Higher Cortical Functions in Man.* New York: Basic Books.

MacLean, C. R. K., K. G. Walton, S. R. Wenneberg, et al. 1994. Altered responses to cortisol, GH, TSH and testosterone to acute stress after four months' practice of transcendental meditation (TM). *Annals of the New York Academy of Sciences* 746:381–384.

MacPhee, R.D.E., ed. 1993. *Primates and Their Relatives in Phylogenetic Perspective.* New York: Plenum Press.

Martin, A., L. G. Ungerleider, and J. Haxby. 2000. Category specificity and the brain: The sensory/motor model of semantic representations of objects. In *The New Cognitive Neurosciences*, ed. Gazzaniga. Cambridge Mass.: MIT Press.

Marzi, C. A. 1986. Transfer of visual information after unilateral input to the brain. *Brain and Cognition* 5:163–173.

Matt, D. C. 1997. *The Essential Kabbalah: The Heart of Jewish Mysticism.* San Francisco: Harper.

McCullough, M. E., K. I. Pargament, and C. E. Thoresen, eds. 2000. *Forgiveness: Theory, Practice, and Research.* New York: Guilford Press.

Mendels, J. 1973. *Biological Psychiatry.* New York: John Wiley & Sons.

Mills, L., and G. B. Rollman. 1980. Hemispheric asymmetry for auditory perception of temporal order. *Neuropsychologia* 18:41–47.

Murphy, G. L., and J. M. Andrew. 1993. The conceptual basis of antonymy and synonymy in adjectives. *Journal of Memory and Language* 32:301–319.

Nebes, R. D., and R. W. Sperry. 1971. Hemispheric disconnection syndrome with cerebral birth injury in the dominant arm area. *Neuropsychologia* 9:249–259.

Neher, A. 1962. A physiological explanation of unusual behavior in ceremonies involving drums. *Human Biology* 34:151–161.

Newberg, A., A. Alavi, M. Baime, P. D. Mozley, and E. d'Aquili. 1997. The measurement of cerebral blood flow during the complex cognitive task of meditation using HMPAO-SPECT imaging. *Journal of Nuclear Medicine* 38:95P.

Newberg, A., A. Alavi, M. Baime, and M. Pourdehnad. 2000. Cerebral blood flow during meditation: Comparison of different cognitive tasks. *European Journal of Nuclear Medicine.*

Newberg, A. B., and E. G. d'Aquili. 1994. The near-death experience as archetype: A model for "prepared" neurocognitive processes. *Anthropology of Consciousness* 5:1–15.

———. 2000. The creative brain/the creative mind. *Zygon* 35:53–68.

Nicholson, R. A. 1963. *The Mystics of Islam.* London: Routledge and Kegan Paul.

Noyes, R. 1980. Attitude change following near-death experiences. *Psychiatry* 43:234–242.

Oman, D., and D. Reed. 1998. Religion and mortality among the community-dwelling elderly. *American Journal of Public Health* 88:1469–75.

Pardo, J. V., P. T. Fox, and M. E. Raichle. 1991. Localization of a human system for sustained attention by positron emission tomography. *Nature* 349:61–64.

Penfield, W., and P. Perot. 1963. The brain's record of auditory and visual experience. *Brain* 86:595–695.

Peng, C. K., J. E. Mietus, Y. Liu, et al. 1999. Exaggerated heart rate oscillations during two meditation techniques. *International Journal of Cardiology* 70:101–107.

Persinger, M. A. 1993. Vectorial cerebral hemisphericity as differential sources for the sensed presence, mystical experiences and religious conversions. *Perceptual and Motor Skills* 76:915–930.

———. 1997. I would kill in God's name: Role of sex, weekly church attendance, report of a religious experience, and limbic lability. *Perceptual and Motor Skills* 85:128–130.

Pesenti, M., M. Thioux, X. Seron, and A. DeVolder. 2000. Neuroanatomical substrates of arabic number processing, numerical comparison, and simple addition: A PET study. *Journal of Cognitive Neuroscience* 12:461–479.

Pinker, S. 1999. *How the Mind Works.* New York: Norton.

Porges, S. W., J. A. Doussard-Roosevelt, and A. K. Maiti. 1994. Vagal tone and the physiological regulation of emotion. *Monographs of the Society for Research in Child Development* 59:167–186.

Posner, M. I., and S. E. Petersen. 1990. The attention system of the human brain. *Annual Review of Neuroscience* 13:25–42.

Posner, M. I., and M. E. Raichle. 1994. *Images of Mind.* New York: Scientific American Library.

Preuss, T. M. 1993. The role of the neurosciences in primate evolutionary biology: Historical commentary and prospectus. In *Primates and Their Relatives in Phylogenetic Perspective*, ed. MacPhee. New York: Plenum Press.

———. 2000. What's human about the human brain? In *The New Cognitive Neurosciences*, ed. Gazzaniga. Cambridge Mass.: MIT Press.

Pribram, K. H. 1981. Emotions. In *Handbook of Clinical Neuropsychology*, eds. Filskov and Boll. New York: Wiley.

Pribram, K. H., and A. R. Luria, ed. 1973. *Psychophysiology of the Frontal Lobes.* New York: Academic Press.

Pribram, K. H., and D. McGuinness. 1975. Arousal, activation, and effort in the control of attention. *Psychological Review* 82:116–149.

Raine, A., M. S. Buchsbaum, J. Stanely, et al. 1994. Selective reductions in prefrontal glucose metabolism in murderers. *Biological Psychiatry* 29:14–25.

Ramachandran, V. S., W. S. Hirstein, K. C. Armel, E. Tecoma, and

V. Iragui. 1997. The neural basis of religious experience. Paper presented at the Annual Conference of the Society of Neuroscience. Abstract #519.1. Vol. 23, Society of Neuroscience.

Reagan, M., ed. 1999. *The Hand of God.* Kansas City: Andrews McMeel Publishing.

Rightmire, G. P. 1984. *Homo sapiens* in Sub-Saharan Africa. In *The Origins of Modern Humans*, eds. Smith and Spencer. New York: Alan R. Liss.

———. 1990. *The Evolution of Homo erectus.* New York: Cambridge University Press.

Ring, K. 1980. *Life at Death: A Scientific Investigation of the Near-Death Experience.* New York: Quill Publishers.

Rothenbuhler, E. W. 1998. *Ritual Communication: From Everyday Conversation to Mediated Ceremony.* Thousand Oaks, Calif.: Sage Publications.

Ryding, E., B. Bradvik, and D. H. Ingvar. 1996. Silent speech activates prefrontal cortical regions asymmetrically, as well as speech-related areas in the dominant hemisphere. *Brain and Language* 52:435–451.

Sagan, C. 1986. *Contact.* New York: Pocket Books.

Saver, J. L., and J. Rabin. 1997. The neural substrates of religious experience. *Journal of Neuropsychiatry and Clinical Neurosciences* 9:498–510.

Savitzky, A. 1999. Cognition, emotion and the brain: A different view. *Medical Hypotheses* 52:357–362.

Schiavetto, A., F. Cortese, and C. Alain. 1999. Global and local processing of musical sequences: An event-related brain potential study. *Neuroreport* 10:2467–2472.

Schrödinger, E. 1964. *My View of the World.* London: Cambridge University Press.

———. 1969. *What Is Life? And Mind and Matter.* London: Cambridge University Press.

Secknus, M. A., and T. H. Marwick. 1997. Evolution of dobutamine echocardiography protocols and indications: Safety and side effects in 3,011 studies over 5 years. *Journal of the American College of Cardiology* 29:1234–1240.

Segal, R. A., ed. 1998. *The Myth and Ritual Theory.* Malden, Mass.: Blackwell Publishers.

Selnes, O. A. 1974. The corpus callosum: Some anatomical and functional considerations with special reference to language. *Brain and Language* 1:111–139.

Shermer, M. 2000. *How We Believe: The Search for God in an Age of Science.* New York: W. H. Freeman & Company.

Shurley, J. 1960. Profound experimental sensory isolation. *American Journal of Psychiatry* 117:539–545.

Singer, R., and J. Wymer, eds. 1982. *The Middle Stone Age at Klasies River Mouth in South Africa.* Chicago: University of Chicago Press.

Smirnov, Y. A. 1989. On the evidence of Neanderthal burial. *Current Anthropology* 30:324.

Smith, B. D., R. Kline, K. Lindgren, M. Ferro, D. A. Smith, and A. Nespor. 1995. The lateralizing processing of affect in emotionally labile extroverts and introverts: Central and autonomic effects. *Biological Psychology* 39:143–157.

Smith, F. H., and F. Spencer, eds. 1984. *The Origins of Modern Humans: A World Survey of the Fossil Evidence.* New York: Alan R. Liss.

Smith, W. J. 1979. Ritual and the ethology of communicating. In *The Spectrum of Ritual*, eds. d'Aquili, Laughlin, and McManus. New York: Columbia University Press.

Spelke, E. S., K. Breinlinger, J. Macomber, and K. Jacobson. 1992. Origins of knowledge. *Psychological Review* 99:605–632.

Sperry, R. 1966. Brain bisection and the neurology of consciousness. In *Brain and Conscious Experience*, ed. Eccles. New York: Springer Verlag.

Sperry, R. W., M. S. Gazzaniga, and J. E. Bogen. 1969. Interhemispheric relationships: The neocortical commissures; syndromes of hemisphere disconnection. In *Handbook of Clinical Neurology*, eds. Vinken and Bruyn. Amsterdam: North Holland Publishing Co.

Sudsuang, R., V. Chentanez, and K. Veluvan. 1991. Effect of Buddhist meditation on serum cortisol and total protein levels, blood pressure, pulse rate, lung volume and reaction time. *Physiology and Behavior* 50, 543–548.

Suedfeld, P. 1964. Conceptual structure and subjective stress in sensory deprivation. *Perceptual and Motor Skills* 19:896–898.

Swisher, L., and I. Hirsch. 1971. Brain damage and the ordering of two temporally successive stimuli. *Neuropsychologia* 10:137–152.

Teasdale, W. 1999. *The Mystic Heart: Discovering a Universal Spirituality in the World's Religions.* Novato, CA: New World Library.

Telles, S., R. Nagarathna, and H. R. Nagendra. 1995. Autonomic chances during "OM" meditation. *Indian Journal of Physiology and Pharmacology* 39:418–420.

———. 1998. Autonomic chances while mentally repeating two syllables—one meaningful and the other neutral. *Indian Journal of Physiology and Pharmacology* 42:57–63.

Turner, V. 1969. *The Ritual Process: Structure and Anti-Structure.* Ithaca, N.Y.: Cornell University Press.

Tuttle, R., ed. 1972. *The Functional and Evolutionary Biology of Primates.* Chicago: Aldine.

Underhill, E. 1990. *Mysticism.* New York: Doubleday.

———. 1999. *The Essentials of Mysticism.* Boston: Oneworld Publications.

Vazquez, M. I., and J. Buceta. 1993. Relaxation therapy in the treatment of bronchial asthma: Effects on basal spirometric values. *Psychotherapy and Psychosomatics* 60:102–112.

Vercelletto, M., M. Ronin, M. Huvet, C. Magne, and J. R. Feve. 1999. Frontal lobe dementia preceding amyotrophic lateral sclerosis: A neuropsychological and SPECT study of five clinical cases. *European Journal of Neurology* 6:295–299.

Venkatesananda, S., tran. 1993. *Vasistha's Yoga.* Albany, N.Y.: State University of New York Press.

Vernet-Maury, E., O. Alaoui-Ismaili, A. Dittmar, G. Dellhomme, and J. Chanel. 1999. Basic emotions induced by odorants: A new approach based on autonomic pattern results. *Journal of the Autonomic Nervous System* 75:176–183.

Vinken, P. J., and C. W. Bruyn, eds. 1969. *Handbook of Clinical Neurology.* Vol. 4. Amsterdam: North Holland Publishing Co.

Walter, V. J., and W. G. Walter. 1949. The central effects of rhythmic sensory stimulation. *Electroencephalography and Clinical Neurophysiology* 1:57–85.

Weingarten, S. M., D. G. Charlow, and E. Holmgren. 1977. The relationship of hallucinations to the depth of structures of the temporal lobe. *Acta Neurochirurgica* 24:199–216.

Weiskrantz, L. 1986. *Blindsight: A Case Study and Implications.* Oxford: Oxford University Press.

———. 1997. *Consciousness Lost and Found: A Neuropsychological Exploration.* New York: Oxford University Press.

Wolford, G., M. B. Miller, and M. Gazzaniga. 2000. The left hemisphere's role in hypothesis formation. *Journal of Neuroscience* 20:RC64.

Worthington, E. L., T. A. Kurusu, M. E. McCullough, and S. J. Sandage. 1996. Empirical research on religion and psychotherapeutic processes and outcomes: A ten-year review and research prospectus. *Psychological Bulletin* 119:448–487.

Zuckerman, M., and N. Cohen. 1964. Sources of reports of visual and auditory sensations in perceptual-isolation experiments. *Psychological Bulletin* 62:1034–1056.

Zukav, G. 1979. *The Dancing Wu Li Masters.* New York: Quill.

INDEX

Absolute Unitary Being, 120, 122, 123, 126–127, 147, 148, 151, 153, 155, 156, 160–162, 171, 172, 198
abstractive operator, 49, 149, 188
access consciousness, 40
acetic acid, 89
acetylcholine, 177
activation studies, 18
active style of meditation, 117, 120–123, 134–135, 164, 196
Adams, Douglas, 142
Adonis, 62
adrenal glands, 38
adrenaline, 38
Alavi, Abass, 174
American Psychiatric Association, 130
amygdala, 43, 44–45, 67, 71, 88–89, 186
Angela of Foligno, 7
animal rituals, 82–83, 85
antelopes, 57
Anweisenheit, 152
Appana samahdi (state of absorption), 41
archetypes, interpretation of, 75
Armstrong, Karen, 56, 104–106, 160, 162–163
arousal system, 38–42, 44, 46, 58–61, 71, 72, 118–119, 121, 125, 183–184
artificial intelligence, 11–14
association areas, 24–25
  attention (AAA), *4, 19, 20,* 29–31, 93, 94, 117–122, 180–181
  orientation (OAA), *4–7, 19,* 20, 28–29, 87, 114–117, 119, 121, 122, 176, 180
  verbal conceptual, *19,* 31–32, 182
  visual, *19,* 26–27

attention association area (AAA), *4, 19,* 20, 29–31, 93, 94, 117–122, 180–181
Austin, James, 185–186
Australopithecus, 65–66
autonomic nervous system, 37–44, 79, 86, 89–90, 183–185

Baime, Michael, 174
binary operator, 50–51, 63–65, 70, 188
Bistami, Abu Yizad, 105
Bittul hayesh (to annihilate the ego), 104
Black Elk, 103
blindsight, 25–26
blood pressure, 8, 38, 39, 44, 86, 129, 131, 184
body temperature, 38, 186
Bohr, Neils, 153
Brahman, 158
Brahman–atman, 120, 147
brain
  activation studies, 18
  arousal system, 38–42, 44, 46, 58–61, 71, 72, 118–119, 121, 125, 183–184
  association areas. *See* association areas
  australopithecine, 65–66
  cerebral cortex, 18–23, 27, 59, 61
  cognitive imperative, 60–61, 64, 67, 190–191
  definition of, 33
  development of self, 149–151
  evolution of, 15–16, 64–66, 177, 178, 191
  functions, 17–18
  hemispheres of, 19–23, 28, 49, 68–73, 179, 188, 192
  Homo erectus, 66, 192
  imaging. *See* brain imaging studies
  limbic system. *See* limbic system
  meditation and. *See* meditation
  Neanderthal, 64
  premotor area of, 92–93
  quiescent system, 38–43, 46, 72, 114, 118–119, 121, 125, 183–184
  robotic, 11–14, 46
  stem, *19,* 43
  structure of, 17
  survival and, 15–17, 38, 168
brain imaging studies, 30, 45
  fMRI (functional magnetic resonance imaging), 53, 175, 180
  PET (positron emission tomography) scan, 53, 175, 181, 182
SPECT (single photon emission computed tomography) scan, 3–5,

brain imaging studies (*cont'd*)
7–9, 34, 36, 53, 126, 146, 175
brain stem, *19*, 43
Brain, Symbol, and Experience (Laughlin, McManus, and d'Aquili), 183
breathing, 38, 39, 104
Bruteau, Beatrice, 167, 168
Buddha, 55, 62, 91
Buddhism, 40, 41, 74–75, 80, 105, 132, 158
butterfly, mating ritual of, 83–85

calcium ions, 177
Calvary Episcopal Church, Pittsburgh, 77
Campbell, Joseph, 55, 56, 74, 75, 91, 92
Cartesian dualism, 183
causal operator, 50, 63, 65, 67–69, 188
cerebellum, 178–179
cerebral cortex, 18–23, 27, 59, 61
chanting, 40, 87, 88, 90, 92, 105
Chargaff, Edwin, 154
chimpanzees, 65
cognitive imperative, 60–61, 64, 67, 190–191
cognitive modules, 187
cognitive operators, 46–53
  abstractive, 49, 149, 188
  binary, 50–51, 63–65, 70, 188
  causal, 50, 63, 65, 67–69, 188
  emotional value, 52–53, 189
  existential, 51–52, 149, 188–189
  holistic, 48, 187
  quantitative, 47, 49, 188
  reductionist, 48, 188
complex partial seizure, 111
concentration, prolonged, 39
Contact (Sagan), 154–155
contemplative prayer, 41, 87, 97, 98, 103, 105, 134–135
Copernicus, 169
corpus callosum, 179
cortisol, 86
courtship rituals, 83–85

Dali Lama, 165, 166
Damasio, Antonio, 52, 183, 189, 194
dancing, 39, 41, 42, 86, 88, 96
d'Aquili, Eugene, 1–10, 82, 183
Darwin, Charles, 169
death
  animal understanding of, 57–58
  certainty of, 61, 132–133
  Neanderthal mortuary rituals, 54–55
déjà vu, 42
dervishes, 91–92
Descartes' Error (Damasio), 183

digestion, 38, 39
Dionysus, 62
disconnection syndrome, 23, 179
doomsday cults, 139
dopamine, 177
dualism, 183

Ebner, Margareta, 98–100, 122
echolalia, 94, 195
echopraxia, 94
Eckhart, Meister, 35–36, 102, 159
Ein-Sof, 159
Einstein, Albert, 15, 153–154, 170
Eleazar, Rabbi, 104
electroencephalography (EEG), 30–31, 196
elephants, 57
emotional brain. *See* limbic system
emotional value operator, 52–53, 189
endocasts, 191, 192
epilepsy, 8, 22, 110–111
Epstein, Perle, 104
Essential Kabbalah, The (Matt), 159
eucharist, sacrament of, 91
Eureka Response, 72
evolution
  of brain, 15–16, 64–66, 123–124, 177, 178, 191
  ritual and, 81–85
existential operator, 51–52, 149, 188–189

'Fana (annihilation), 105
fasting, 88, 103, 105, 193
fast rituals, 115, 138, 196
Feeling of What Happens, The (Damasio), 52
Feuerbach, Ludwig, 128
fight-or-flight response, 38, 58
flat worms, 15, 16, 177–178
fMRI (functional magnetic resonance imaging), 53, 175, 180
"Footnote to All Prayers, A" (Lewis), 157–158, 160, 162, 171
fragrances, 88, 89, 193
Franciscan nuns, 7, 36
Frazer, James, 128
Freud, Sigmund, 99, 107–108, 128, 130, 145, 195
frontal lobe, *19,* 20, 30, 176, 178, 181–182

Gage, Phineas, 182
Galileo, 169
Gould, Stephen Jay, 124
Greek mysticism, 104–105
Greeley, Andrew, 107, 108
greeting rituals, 83
Gregorian chants, 82
grooming rituals, 83

group prayer, 40
growth hormone, 44, 186

Hallaj Husain ibn Mansur, 102
hallucinations, 8, 42, 107, 109, 111–113, 189
Hawking, Stephen, 15
health benefits of religion, 129–131
heart rate, 38, 39, 86, 131, 184–185, 193
hesychia (inner silence), 104
Hinduism, 6–7, 158
hippocampus, 43, 45–46, 67, 86–87, 114, 117, 118, 121, 178, 186, 194
History of God, A (Armstrong), 56, 104–106, 162–163
holistic operator, 48, 187
Homo erectus, 66, 192
hormones, 44, 186
Huang Po, 147–148
hunger, 186
hyperarousal, 40–42
hyperquiescence, 40, 41
hyperventilation, 88, 193
hypothalamus, 43–44, 72, 86, 89, 118, 119, 121, 125, 178, 186

illusions, 42
immune system function, 86, 129, 131, 186
incense, 88, 89
infants, mathematical concepts and, 49, 188
inferior parietal lobe, 51, 149, 192
intentionality, 152
interspirituality, 166
Islam, 91, 158, 159

James, William, 106–107
Jesus, 55, 56, 62, 74, 91
Jewish mysticism, 103, 104
Joan of Arc, 111
Joseph, Rhawn, 180, 181, 185
Jung, Carl, 75, 153

Kabbalah (Epstein), 104
Kabbalistic mysticism, 103, 104, 158–159
Koenig, Harold, 129–130
Kuwamara, Leslie, 151

Landau, Misia, 191
language, 20–22, 31
Lao-tzu, 102
latah, 93–94
Laughlin, Charles, 82, 183
lavender, 89
left brain hemisphere, 19–23, 28, 49, 68–71, 73, 179, 188
Lewis, C. S., 157–158, 160, 162, 171

Lilly, John, 153
limbic system, 30, 37, 42–46, 52, 182, 189
  amygdala, 43, 44–45, 67, 71, 88–89, 186
  of animals, 58
  hippocampus, 43, 45–46, 67, 86–87, 114, 117, 118, 121, 178, 186, 194
  hypothalamus, 43–44, 72, 86, 89, 118, 119, 121, 125, 178, 186
Li Po, 151
long-distance swimmers, 41
Lyell, Charles, 169

McManus, John, 82, 183
mantra, 90, 117
marathon runners, 41
marked actions, 88
Marx, Karl, 128
mass, ritual of the, 95
material reality, 143–145, 152, 155, 170, 183, 198
mathematical operations, 49, 188
Matt, Daniel, 159
Mayan pyramids, 75
meditation, 1–8, 86, 97, 104, 173–176
  active approach, 117, 120–123, 134–135, 164, 196
  autonomic nervous system and, 39–41, 185
  binary operator, 50–51, 188
  EEG readings during, 30–31, 196
  heart rate during, 185, 193
  hormone release and, 44
  limbic system and, 44
  passive approach, 117–120, 196
  quantitative operator, 47, 49, 188
  visions during, 27
mental health, mysticism and, 107–113
mind, definition of, 33
Mohammed, 111
mortification, 103
mortuary rituals, Neanderthal, 54–55
multimodal areas, 180
multiple sclerosis, 177
music, 77–80, 114–115, 146, 192
myelin cells, 177
Mystic Heart, The (Teasdale), 135, 161–162, 165–167
mysticism, 98–127, 171
  defined, 100–107
  Freudian explanation of, 99, 107–108
  Kabbalistic, 103, 104, 158–159
  and mental health, 107–113
  neurobiology of mystical experience, 113–127

mysticism (*cont'd*)
origins of religion and, 133–140
reality and, 142–145, 151–153, 155–156
Mysticism (Underhill), 35, 100
myths, 8, 9, 54–76, 171
Buddha, 55, 62
cognitive operators and, 63–64
cultural similarities in, 74–75
framework of, 62–63, 191
genesis of, 64–76
Jesus, 55, 56, 62, 74
meaning of term, 56
-ritual connection, 9, 32, 90–96

National Opinion Research Center, 107
Native Americans, 96, 103
natural selection, 124, 132
Neanderthals, 132
brain of, 64, 66
mortuary rituals, 54–55
near-death experiences, 8, 27, 183
neocortex, 19, 43, 44, 178
nerve cells, 15, 16, 177, 179
neurons, 24, 33, 179–180
neurotransmitters, 24, 177
Nietzsche, Friedrich, 128–129, 169
Nirvana, 120, 147

objective reality, 143–145, 152, 155, 170, 198
occipital lobe, 17, *19,* 20, 31
olfactory system, 89, 194
One Mind, 147–148
Oppenheimer, Robert, 153
orgasm, 125
orientation association area (OAA), 4–7, *19, 20,* 28–29, 87, 114–117, 119, 121, 122, 176, 180
Origin of Species, The (Darwin), 169
Osiris, 62
out-of-body experiences, 42, 110, 183

parasympathetic nervous system. *See* quiescent system
parietal lobe, 4, *19,* 20, 28, 31, 65–66, 188
passive style of meditation, 117–120, 196
patriotic rituals, 90
peak state, 31, 173–174. *See also* mysticism
Persinger, Michael, 182–183, 185
personalized God, 160–163

PET (positron emission tomography) scan, 53, 175, 181, 182
Phantasia catalyptica, 152
Polynesian fertility dance, 82
prayer, 86, 90
  contemplative, 41, 87, 97, 98, 103, 105, 134–135
  healing powers of, 8
prefrontal cortex, 29, 93, 180
premotor area of brain, 92–93
primary receptive areas, 24
primary vision area, *20*
primates, 65, 66, 192
Principles of Geology (Lyell), 169
psychotics, 109–110

quantitative operator, 47, 49, 188
quantum mechanics, 153
quiescent system, 38–43, 46, 72, 114, 118–119, 121, 125, 183–184

Rabi'a al-Adawiyya, 159
Rabin, John, 101, 109, 110
Ramachandran, V. S., 32, 185
reality, experience of, 35–37, 142–156, 174
reductionist operator, 48, 188
redundancy, 29–30
reification, process of, 149–150
relativity, theory of, 153
relaxation, 38–40
religion, 86. *See also* mysticism
  contemplative prayer, 41, 87, 97, 98, 103, 105, 134–135
  health benefits of, 129–131
  origins of, 133–140
  religious harmony, 165–166
  religious intolerance, 163, 164
  sense of control and, 131–133, 164
religious awe, 89, 110
repetition, in ritual, 82, 85, 88, 97, 193
reproductive hormones, 186
Restaurant at the End of the Universe, The (Adams), 142
reticular activating system (RAS), 178
rhythm, ritual and, 77–80, 82, 85–89, 97, 138, 192–193
right brain hemisphere, 19–23, 28, 69, 70, 72, 73, 179, 188, 192
ritual, 8–9, 77–97, 171
  animal, 82–83, 85
  and emotion, 88–90
  evolutionary roots of, 81–85
  -myth connection, 9, 32, 90–96
  neurobiology of, 86–90
  rhythm and repetitive

ritual (*cont'd*)
behavior and, 77–80, 82, 85–89, 97, 192–193
secular, 81, 90
and unity, 80–81
robotics, 11–14, 46
Roman Catholic Church, 91, 95
rosary groups, 96
running, 39, 41
Russell, Bertrand, 128

sacrificial gifts, 137
Sagan, Carl, 154–155
St. Paul, 111
Saver, Jeffrey, 101, 109, 110
schizophrenia, 8, 107, 109
Schrödinger, Edwin, 153, 154
secondary receptive areas, 24
secular rituals, 81, 90
seizures, 110–112
self, neural development of, 149–151
self-transcendence, 79, 80, 115
septal nucleus, 178
sexual response, 42, 125–126, 139
Shakespeare, William, 15
Siddhartha, 74, 155–156
silence, 103
silver-washed fritillary, mating ritual of, 83–85
skin, electrical conductance of, 40
sleep regulation, 38
slow rituals, 115, 138, 196
Smith, Joseph, 111
sodium ions, 177
somatic marker hypothesis, 52, 192, 194
SPECT (single photon emission computed tomography) scan, 3–5, 7–9, 34, 36, 53, 126, 146, 175
Sperry, Roger, 22–23
spinal cord, 19, 178
split-brain patients, 22–23
Stoics, 152
subcortical structures, 19, 178
subjective reality, 144, 145, 152, 155, 198
submission, rituals of, 83
Sufis, 41, 86, 105, 106, 135
Sumerian ziggurats, 75
survival, 15–17, 38, 168
Swedenborg, Emanuel, 111
swimming, 41
sympathetic nervous system. *See* arousal system

Tammuz, 62
tantric yoga, 39
Tao, the, 120, 147
Taoism, 103, 158
Tauler, John, 101
Teasdale, Wayne, 135, 161–162, 165–166
temporal lobe, 17, *19*, 20, 31–32, 44, 45, 110, 185
seizures, 182–183

Teresa of Avila, St., 111–113
testosterone, 44
thalamus, 46, 117, 178, 185
thirst, 186
thyroid stimulating hormone, 44, 186
Tibetan meditation, 1–7, 36, 197
Tourette, Gilles de la, 93–94
trance states, 116
transcendental meditation, 39

Underhill, Evelyn, 35, 100–101, 160
unimodal areas, 180
Unio Mystica, 80, 122, 147, 164
unitary continuum, 115–116
unknowableness of God, 158–161
Upacara samadhi (access consciousness), 40
Upanishads, 6–7

Van Gogh, Vincent, 111
Varieties of Religious Experience (James), 106
vasopressin, 44
verbal conceptual association area, *19,* 31–32, 182
virtual reality, 13
vision quests, 96
visual association area, *19,* 26–27
visual imagery, 27, 94, 195
visual processing system, 25–27
voice recognition, 13
Void Consciousness, 120
Voudon practitioners, 41, 86

Wernicke's area, 192
whole-brain harmony, 73
will, 29–30, 181
Wilson, E. O., 191
worms, 15, 16, 177–178

yoga, 196

Zen and the Brain (Austin), 185
Zen practitioners, 30, 103